DATE DUE

DEC 20 '00			

DEMCO 38-296

INCAPACITATION

STUDIES IN CRIME AND PUBLIC POLICY
Michael Tonry and Norval Morris, *General Editors*

Police for the Future
David H. Bayley

Incapacitation: Penal Confinement and the Restraint of Crime
Franklin E. Zimring and Gordon Hawkins

R

INCAPACITATION

Penal Confinement and the Restraint of Crime

Franklin E. Zimring
Gordon Hawkins

New York   Oxford
OXFORD UNIVERSITY PRESS
1995

An Earl Warren Legal Institute Study

Oxford University Press

Oxford   New York
Athens   Auckland   Bangkok   Bombay
Calcutta   Cape Town   Dar es Salaam   Delhi
Florence   Hong Kong   Istanbul   Karachi
Kuala Lumpur   Madras   Madrid   Melbourne
Mexico City   Nairobi   Paris   Singapore
Taipei   Tokyo   Toronto

and associated companies in
Berlin   Ibadan

Copyright © 1995 by Oxford University Press Inc.

Published by Oxford University Press, Inc.
200 Madison Avenue, New York, New York 10016

Oxford is a registered trademark of Oxford University Press

Library of Congress Cataloging-in-Publication Data
Zimring, Franklin E.
Incapacitation : penal confinement and the restraint of crime /
Franklin E. Zimring and Gordon Hawkins.
p.   cm.—(Studies in crime and public policy)
Includes index.
ISBN 0-19-509233-3
1. Imprisonment—United States.   I. Hawkins, G. J. (Gordon J.)
II. Title.   III. Series.
HV9471.Z55   1994
365'.973—dc20   94-8630

1 3 5 7 9 8 6 4 2
Printed in the United States of America
on acid-free paper

Preface

Of all the justifications for criminal punishment, the desire to incapacitate is the least complicated, the least studied, and often the most important. The major institutions of criminal punishment in the Western world—the prison and the jail—are designed and operated to restrain those under their control. All of the other objectives of incarceration are ancillary to the basic structure of the modern prison and jail: incapacitation is central.

Although it has always been an essential function of the prison and jail, the restraint of offenders is an aspect of incarceration that has rarely been the subject of scholarly attention or scientific assessment. For centuries the disjunction between practical significance and scholarly attention has been a chronic element of criminological theory and practice. Thus incapacitation has been the major motive of incarceration for many decades but has received scant attention in criminology, in criminal law, or in jurisprudence. This function of confinement is too obvious to be visible or important to scholars.

The past two decades have witnessed a modest increase of academic interest in incapacitation, together with a phenomenal growth in both prison populations and in incapacitation as the primary justification of imprisonment. A handful of studies have been published on the scientific basis for incapacitation, but the gulf between what is known about incapacitation and what is claimed for it has never been greater than in the early 1990s. Policy debates on incapacitation are infected with impossible claims. In the summer of 1993, United States Senator Phil Gramm asserted that each year of confinement in the United States saved $430,000 in crime (Gramm, 1993). With 700,000 additional persons confined in the United States since 1980, Senator Gramm's calculations should have produced a $300 billion savings in the cost of crime, an amount greater than the federal deficit or the national defense budget. The senator did not mention where these savings could be found. Perhaps it was an unnecessary detail. In the current debate, the expert on penal policy often seems to be the one who shouts the loudest.

The long tradition of scholarly neglect of this topic is a puzzle worthy of more attention than it will receive in these pages. Perhaps the cardinal defect of incapacitation as an academic research topic is its obviousness. If the criminal justice system locks up 10,000 convicted offenders, it follows as a matter of course that some crimes will be prevented as a consequence. The only significant question relates to the amount of crime that will be prevented. Banal details of this sort are not the kind of material to engage theoretical minds. The services of an engineer or an accountant might seem to be what is required to deal with that sort of question. So the simplicity of the process serves as a disincentive to scholarly attention.

In fact, however, the unanswered questions about incapacitation in criminal justice are among the most intellectually challenging as well as among the most important to be encountered in the social science of crime. Theoretical and methodological problems of great seriousness exist, but can only be discovered by those who direct sustained attention toward the topic.

This book aims to provide a comprehensive survey of the conceptual, methodological, and policy dimensions of incapacitation. We hope to supply some organization for the major conceptual issues, to offer an assessment of the research methods available to study incapacitation, to contribute some original research, and to explore the linkages between knowledge about incapacitation and imprisonment policy. The book is organized in three parts: part I deals with the principal conceptual issues; part II is concerned with research; and part III discusses the links between incapacitation and criminal justice policy.

Our treatment of conceptual issues unfolds in four installments. Chapter 1 provides a history of how the incapacitation of offenders became the dominant official justification for imprisonment in the United States in the 1970s and 1980s. The fact that incapacitation rose to preeminence chiefly by a process of elimination in which other purposes of imprisonment fell out of favor is an important explanation of the lack of evidential support for, and academic attention to, incapacitation. Restraint became the paramount penal purpose in the United States while scholars of criminal law and criminology were looking elsewhere. Indeed, the social scientists' undermining of other purposes of imprisonment facilitated the pattern of dominance by default.

Chapter 2 summarizes the literature on incapacitation from its first analytic discussion by Jeremy Bentham in 1802 through to the studies and commentaries of the mid-1980s. The most striking feature of the survey is that the English-language literature on the topic is so sparse that we can summarize it in one modestly sized book chapter. Most recent publications on the topic conform to a dialectical pattern in which particular claims are made, then met with refutation, and thereafter the specific subject of the controversy disappears from published view.

Chapter 3 discusses three important components of any acceptable theory of incapacitation. The first section, building on existing literature, outlines the way in which it is presumed that when individuals disposed to commit crime are incarcerated, the result will be less crime in the community because those individuals are restrained. A second section of the chapter discusses how variations in individual propensities to commit crime interact with the presence or absence of selectivity in the criminal justice processes that determine which offenders are sentenced to imprisonment. This section makes predictions about the diminishing marginal return from increasing imprisonment in criminal justice systems with selective imprisonment policies. The final section of chapter 3 deals with the relationship between the individual criminal propensities of the persons restrained and the extent of crime reduction experienced in the communities from which they are removed. The lack of one-to-one correspondence between individual and community effects is principally a function of group criminality and the tendency for unexploited criminal opportunities to encourage higher levels of activity among those not restrained.

Chapter 4 deals with the jurisprudence of incapacitation, contrasting the rather extensive analysis of the problems of special incapacitation with the lack of analysis of collective incapacitation. The chapter suggests two limiting principles as ethical constraints on the use of incapacitative imprisonment. We doubt, however, that those limits will function as a sufficient safeguard against excessive incapacitative imprisonment.

The second part of the book, dealing with research, is evenly divided between critical and constructive segments. Chapter 5 examines the most important research strategies available for incapacitation research. The first two of these, offender surveys and official-record studies, are complementary methods of determining the individual-level effects of incarceration. The survey method tends to overestimate crimes prevented by incapacitation, whereas official-record studies of arrest and conviction use raw data that underestimate the criminality of subjects and the extent of the preventive effects of incapacitation. There is no convenient way of averaging these errors and correcting the exaggerations in both methods. The third section of chapter 5 discusses the kind of community-level analysis that has not been done to date but that will be necessary to assess accurately the extent of group criminality and offense substitution.

Chapter 6 recounts our analysis of the effects in California of an increase of more than 100,000 in prison and jail population on the crime rate during the 1980s. Most of our statistical measures suggest a reduction in reported index felony crime of about 3.5 per additional year of confinement with 90 percent of that reduction clustered in burglary and larceny. However, an analysis of trends in arrest by age between 1980 and 1990 casts doubt on incapacitation as the primary cause of the decrease in burglary and theft because arrest rates drop most for juvenile offenders, whereas the increased incapacitation occurs principally among older offenders.

The third part of this book presents two chapters that deal with the

relationship between knowledge regarding incapacitation and the for-
mulation of penal policy. Chapter 7 deals with attempts to measure both
the costs of crime and the costs of law enforcement in dollar terms and
to translate decisions about the desirability or appropriateness of crime
countermeasures into mechanical cost-benefit calculations. The central
problem here is that the economist lacks a concept of the cost of crime
that is both relevant and rigorous. The rigorous concept of efficiency
cost is not relevant to many of the harms that lead to the classification of
human behavior as criminal. Any less precise definition of cost has none
of the coherence and rigor required for economic analysis.

Chapter 8 surveys the relationship between incapacitation as a justification for punishment, and imprisonment policy. It is an attempt to
place incapacitation in the context of other purposes of punishment
and other methods of crime control. A major theme of the chapter is
the contrast between methods of monitoring behavior and attempts to
control behavior by restraint when dealing with populations at risk of
future offending. The chapter also includes discussion about how
trends in imprisonment rates may shift the emphasis back from collective to selective incapacitation in the near future.

The project that produced this book grew out of two long-standing
research interests of its authors. For some years we have been jointly
patrolling the border between criminal law and criminology, paying
particular attention to the relationship between the announced purposes and behavioral impact of criminal punishments. In this broad
sense the current study is part of the same undertaking that led to our
book on *Deterrence* (Zimring and Hawkins, 1973).

However, the immediate stimulus for this project was a series of
studies of prison population in the United States. In 1991 we wrote *The
Scale of Imprisonment* (Zimring and Hawkins, 1991), which was specifi-
cally concerned with the growth of prison population in recent years.
That volume led us to a study of the explosive growth of prison and jail
populations in the state of California, *Prison Population and Criminal Justice Policy in California* (Zimring and Hawkins, 1992). The California
prison and jail statistics developed in that study documented a huge
change in imprisonment policy meant to reduce crime through imprisonment. We then began a project to study the impact of the shift in
incarceration on rates of crime. The current book crystalized around
the need to organize and articulate the major issues concerning the
restraining influence of penal confinement.

Conspicuously lacking from this book are global estimates of the
amount of crime prevention attributable to incapacitation, and across-
the-board judgments about incapacitation as a purpose of imprisonment. Currently available studies do not provide a narrow range of incapacitation estimates, not merely because the studies are preliminary and
inexact but because no narrow range of incapacitation is there to be
found. The amount of crime prevented by incapacitation is both vari-

able and contingent, varying in relation to different social circumstances and under different criminal justice policies. No prospect exists of discovering a unitary level or pattern of crime prevention that might be achieved by penal restraint.

Moreover, the topic of incapacitation as a purpose of penal confinement is not well suited to either unqualified acceptance or total rejection. To discuss imprisonment at all but exclude incapacitation as one of its purposes would be absurd. If prisons are good for anything it is as institutions of restraint. Yet incapacitation is impossibly open-ended as a general principle of criminal punishment. If persons who present some threat of future crime are to be confined, why not confine all of them indefinitely? The balance of desert and proportionality with preventive potential, as well as rough calculations of cost and benefit—these are the inescapable elements of decisions about the scope of preventive confinement in modern criminal justice.

Acknowledgments

The considerable financial and personal support required to complete this volume came from a wide variety of sources. Funding for our research efforts from 1990 to 1993 was provided by the Criminal Justice Research Program of the Earl Warren Legal Institute at the University of California at Berkeley and the Boalt Hall Fund at the School of Law. The California Policy Seminar also provided funding for the research project that is the subject of chapter 6. Holly Brown and Andres Jimenez of the Seminar provided both materiel and moral support.

We were fortunate to have Hank Ibser, a graduate student in statistics at the University of California as our principal statistical resource. David Kupferschmidt of Boalt Hall and Richard Leo and Sam Kamin of the Jurisprudence and Social Policy program at Berkeley provided general research assistance of high quality with great good humor. Materials and references came from Jacqueline Cohen, Mark Mauer, William Spelman, Peter W. Greenwood, and Joan Petersilia.

Everything that happens at the Earl Warren Legal Institute is Karen Chin's responsibility: this book was no exception. She not only presided over the administration of the project but also produced drafts of five chapters of the manuscript. Dong Yun joined us in May 1993 and took a primary role in the completion of the manuscript.

The book was much improved by the critique of draft chapters by Philip J. Cook, Norval Morris, Peter Greenwood, Daniel Nagin, José A. Canela-Cacho, Sheldon Messinger, and Michael Tonry. The California materials included in the first five sections of chapter 6 were presented at the Neyman seminar of the University of California at Berkeley Department of Statistics and also to the Boalt Hall Faculty Colloquium in the spring of 1993.

Berkeley, California   F.E.Z.
June, 1994   G.H.

Contents

I

CONCEPTS

1

Dominance by Default

Incapacitation now serves as the principal justification for imprisonment in American criminal justice: offenders are imprisoned in the United States to restrain them physically from offending again while they are confined. The singular importance of incapacitation as a purpose of imprisonment is of relatively recent vintage. In the 1970s, the rhetoric of rehabilitation was a dominant feature of the literature and discussion of imprisonment, and the deterrence justification was more prominent than incapacitation in debates about punishment. It is only in the last fifteen years that something approaching a consensus about the priority of restraint has begun to emerge.

Not only is the dominance of incapacitation a recent phenomenon, it is based on almost unexamined principles. Neither the arguments in support of using incapacitation simply to restrain the offender nor the evidence to support the proposition that such incapacitation lowers community crime rates has been subjected to careful scrutiny or detailed analysis.

The amount of scholarly attention devoted to the incapacitation process has been minimal. Theories regarding incapacitative effects have been published only sporadically. Almost all the statistical arguments regarding incapacitation have relied on casually assembled secondary statistics relating to crime and imprisonment. Original data have been gathered for only a handful of studies, and there is no indication that this deficit in public attention is likely to be addressed in the near future by an increase either in the priority given to questions about the nature and limits of incapacitative effects or in the resources devoted to empirical research on the subject.

We think the unexamined character of incapacitation doctrine is both regrettable and explicable. We argue in this chapter that much of the explanation for the lack of scrutiny of imprisonment as restraint derives from the way in which this particular motive for imprisonment ascended to its present prominent position. This did not come about as a result of either scholarly or public attention devoted to the preventive effect of locking people up or to any significant featuring of this topic in political or academic debate about the purposes of imprisonment. Instead, incapacitation rose to prominence by a process of elimination as scholarly and public debate about *other* functions of imprisonment undermined faith in prison rehabilitation as an effective process and in deterrence as a basis for making fine-tuned allocations of imprisonment resources. Attention during the period when incapacitation rose to prominence was on other issues so that the victory secured by incapacitation was won by default. Yet the unexamined status of a central tenet of the justification for the incarceration of more than a million citizens is an intolerable feature of the administration of criminal justice in the United States.

This chapter analyzes the correctional debate in the American criminal justice system after the 1950s. Arguments about rehabilitation laid the groundwork for the emergence of incapacitation as a residual justification for imprisonment. Section I examines the need for theories of imprisonment to support the practice and reassure those who participate in it. Section II shows how the decline of faith in prison as an instrument of rehabilitation left imprisonment as a practice in search of a theory. Section III argues that both liberal and conservative critics of rehabilitation imagined incapacitation as the proper residual function of imprisonment, but with sharply different notions of the residual function's size and defining characteristics. Many liberals imagined mini-prisons relegated to the confinement of small numbers of habitual offenders or specially dangerous persons, whereas many conservatives envisaged general policies of incapacitation broad enough to maintain and expand prison numbers. So the debate was between two versions of incapacitation logic—between more selective and more general incapacitation—rather than between incapacitation and competitive aims. In this competition, the liberal conception was burdened thrice: by public concern about crime, by the absence of credible nonprison sanctions for serious crime, and by the way that special predictions of dangerousness required some of the same procedures and judgments that had been tarnished in rehabilitative regimes. Since the debate was between two versions of a restraint rationale, the contending parties did not question any basic assumptions about incapacitation as an animating policy.

Section IV contrasts the policy focus on incapacitation with the continued dominance of rehabilitation and inattention to restraint in academic circles. Section V discusses the rhetorical advantages of incapacitation over desert and deterrence, two possible successors to rehabilitation as a justification for prison. Finally, section VÍ demonstrates the role of incapacitation rhetoric in the dialogue about expanding prison systems.

I. Paradigms of Imprisonment

There is no official or single source that one can look to for the justification of the prison enterprise at any given time. Imprisonment is justified or explained in a variety of different contexts, including public and judicial discussion of prison as a criminal sanction and political debates about the building and funding of prisons. Indeed, at the outset the question arises whether or why any justification for imprisonment is necessary.

In a formal or legal sense, no need may exist for any particular explanation—which would be the equivalent of a jurisprudence of imprisonment—about the use of prisons to punish crime, as opposed to other types of sanctions. Yet those who administer prisons, those who build them, and those who are officially involved in sending convicted offenders to prison, do feel the need to refer to the purposes of imprisonment in the course of carrying out their various duties. The ways in which imprisonment is justified change significantly over time.

The sources we look to in seeking the justification of imprisonment include political debates, academic penology, and moral and political philosophy, as well as the pronouncements of judicial officers in the context of criminal sentencing. Although, it is logically and legally possible to continue both to administer prisons and to use imprisonment as a punishment without the support of any specific justification or ideology of imprisonment, it would be difficult in a political democracy to do so without any positive sense of purpose or function for them. Those who work in prison, those who sentence offenders to prison, and those who support the institution in less palpable ways all need some paradigm of imprisonment, a sharp image of what prisons are needed for and may achieve.

The need to justify punishment is reflected in moral logic as well as history. Since penal practices are by definition unpleasant, the world is a poorer place for their presence unless the positive functions achieved by them outweigh the negative elements inherent in the policies. "But all punishment is mischief: All punishment in itself is evil," said Bentham, and the felt need to counterbalance the mischief that prison does is always present (1843:83).

The history of the use of imprisonment as a legal punishment illustrates the perennial nature of the demand for justification. One of the earliest historical references to imprisonment, in the Emperor Justinian's *Digest*, is a prohibitory injunction. In the *Digest*, the Emperor refers to a principle coined by the Roman jurist Ulpian: "Carcer ad continendos homines non ad puniendos haberi debet [Prison ought to be used for detention only, but not for punishment]." This reference to Ulpian's principle in effect enacted it as imperial law and rendered imprisonment illegal as a punishment (Grünhut, 1948:11).[1]

So for a thousand years after Justinian had adopted this principle (in 533 A.D.) the penal law of Europe was, in theory at any rate, dominated by the notion of the illegality of imprisonment as a punishment. The function of prisons or jails was merely to hold offenders until *punishment,* in the form of either death, banishment, transportation, the forfeiture of property, or a variety of corporal penalties, could be enforced or inflicted. Yet although in principle Justinian's rule prevailed, there is, as Sean McConville has demonstrated, "abundant confirming evidence that imprisonment was used punitively in medieval times" (McConville, 1981:2; see also Bassett, 1943, and Pugh, 1968). "This ambiguity between administrative practice and legal doctrine," says Grünhut, "prevailed throughout the middle ages and up to modern times" (1948:11). It is not surprising, therefore, that from the earliest days the use of imprisonment as a legal punishment has required justification.

Some conception of what prisons are good for is a political and cultural necessity, even if it is not a strict requirement for a criminal law that uses imprisonment as a sanction. Indeed, the fall from favor of a particular justification of imprisonment puts pressure on many actors in the system to find alternative acceptable rationales. The decline of the rehabilitative ideal as a paradigm justifying imprisonment created the vacuum that incapacitation has so quickly filled.

II.   The Debate about Rehabilitation

It would be difficult to overstate the degree to which the concepts and vocabulary of rehabilitation have dominated discourse about the purposes and functions of imprisonment in modern American history. The professional field concerned with the administration of prisons and jails is called corrections. Institutions of confinement are referred to as reformatories, training schools, and correctional institutions. According to the rehabilitation ideal, not only is the reform, reorientation, and rehabilitation of the convicted offender seen as the official purpose of the

---

[1]The text in the *Digest* (D. 48.19.8.9) says that governors were in the habit of condemning people to be kept in prisons or in chains; that they ought not to do so; that it is not an authorized punishment; that prison is for confining people who are awaiting trial or who are to be examined. The most commonly used authorized punishments for serious offenses were, for those of high social status and with means to live, exile either for a limited period or permanently; for those of low social status and slaves, forced labor. (We are indebted to John H. Langbein, Chancellor Kent Professor of Law and Legal History at Yale Law School, and Peter G. Stein, Regius Professor of Civil Law at the University of Cambridge, for this information.)

prison sentence but judgments about progress in rehabilitation programs are supposed to provide the basis for determining when sentences should be terminated in favor of parole to the community. For most of the twentieth century, the concept of rehabilitation has dominated penal policy and practice by acclamation and largely without dissent.

Rehabilitation was the law's stated objective in the criminal sentencing system and remained the dominant ideology in the architecture of the model penal code reforms of the 1960s. Correctional administrators were no less uniformly enthusiastic about reform as a penal purpose. Even reformers, who were engaged in the task of criticizing and pointing out the faults of the American prison establishment, disagreed for the most part about the institutional means rather than the ends of American prison and correctional programs.

Two passages from *The Challenge of Crime in a Free Society,* the general report of the 1967 President's Commission on Law Enforcement and Administration of Justice, are relevant and representative:

> Life in many institutions is at best barren and futile, at worst unspeakably brutal and degrading. To be sure, the offenders in such institutions are incapacitated from committing further crimes while serving their sentences, but the conditions in which they live are the poorest possible preparation for their successful reentry into society, and often merely reinforce in them a pattern of manipulation or destructiveness.

These deplorable conditions, however, were not a basis for concluding the rehabilitative ideal was wrong but instead meant that conditions should be improved so that the system could live up to its rehabilitative premises:

> However, there are hopeful signs that far-reaching changes can be made in present conditions. The Commission found, in the course of its work, a number of imaginative and dedicated people at work in corrections. It found a few systems where their impact, and enlightened judicial and legislative correctional policies, had already made a marked difference; a few experimental programs whose results in terms of reduced redicivism were dramatic. A start has been made in developing methods of classification that will permit more discriminating selection of techniques to treat particular types of offenders. But many of the new ideas, while supported by logic and some experience, are yet to be scientifically evaluated. Nevertheless, the potential for change is great. (President's Commission, 1967:159)

These passages are typical of the attitudes and orientation of social welfare liberals in the late 1960s, when reform was generally accepted by prison administrators as the purpose of and the justification for imprisonment.

Reformation owed its uniform popularity in large part to the optimistic story it told about official motives for the uses of imprisonment and the objectives of those who administered penal institutions. The ideology of rehabilitation regarded the imprisoner and the imprisoned as collaborators in a process designed to improve the prisoner's prospects. Viewed in cynical retrospect, it may call to mind the old joke about the world's three greatest lies, one of which ran, "I'm from the government and I'm here to help you." Yet the sincerity of many adherents to the rehabilitative idea cannot be doubted even by the harshest of revisionist historians.

We do not intend to give a detailed history of the erosion of public confidence in rehabilitation as the major purpose of imprisonment, as expounded in Francis Allen's excellent study *The Decline of the Rehabilitative Ideal,* based on his Storrs Lectures on Jurisprudence at Yale Law School in 1979. Allen defines the rehabilitative ideal as

> the notion that a primary purpose of penal treatment is to effect changes in the characters, attitudes, and behavior of convicted offenders, so as to strengthen the social defense against unwarranted behavior, but also to contribute to the welfare and satisfaction of offenders.

Though some criticisms of, or reactions to, the ideal "comprise essentially irritated responses to the prevalence of crime and offer only an all-encompassing faith in the efficacy of coercion and repression," Allen demonstrates that the challenge to the rehabilitative ideal is not unsupported by serious considerations.

He describes the three principal propositions that embody those considerations:

> First, the rehabilitative ideal constitutes a threat to the political values of free societies. Second—a distinct but closely related point—the rehabilitative ideal has revealed itself in practice to be peculiarly vulnerable to debasement and the serving of unintended and unexpressed social ends. Third, either because of scientific ignorance or institutional incapacities, a rehabilitative technique is lacking; we do not know how to prevent criminal recidivism by changing the characters and behavior of offenders.

He continues:

> In a remarkably short time a new orthodoxy has been established asserting that rehabilitative objectives are largely unattainable and that rehabilitative programs and research are dubious or misdirected. The new attitudes resemble in their dominance and pervasiveness those of the old orthodoxy, prevailing only a few years ago, that mandated rehabilitative efforts and exuded optimism about rehabilitative capabilities. (Allen, 1981:2, 33–34, 547–58)

Whether the change is best characterized as a gradual decline or a sudden collapse, the extent of and the timing of the shift away from rehabilitation as the principal justification for imprisonment seem clear.

Before the late 1960s the disappointing results of correctional treatment were regarded as a challenge—a mandate for further experiment or changing treatment approaches. Whatever the deficiencies of particular treatment or correctional programs, the ideal of treatment as the principal justification for imprisonment remained undisturbed. Yet by the late 1970s, public confidence in rehabilitation had been thoroughly undermined at least with respect to adult offenders. Some landmarks in this radical turnaround include Robert Martinson's celebrated article "What Works? Questions and Answers about Prison Reform" (Martinson, 1974); the passage of the California Determinate Sentencing Legislation in 1976, with its explicit rejection of rehabilitation as a motive of imprisonment; and the report of the National Academy of Sciences Panel on Rehabilitation in 1979 (Sechrest et al., 1979).

Three aspects of this sharp reversal of fortune for the rehabilitative ideal deserve special mention. First, the deflation of that ideal happened over a relatively short period of time without any major new empirical finding or scientific theory to initiate or support it. Robert Martinson and his associates had assembled extensive material documenting the lack of success in correctional treatment of programs, but these negative findings antedated the successful attack on rehabilitation by many years (Lipton et al., 1975).

Second, the normally warring ideologies that pervade the American criminal justice policy debate were united in hostility to rehabilitation by the mid-1970s just as they had been united in support of rehabilitation a very few years before. The rejection of rehabilitation in the provisions of the 1976 California Penal Code was enthusiastically supported by a coalition that included police chiefs, district attorneys, Quakers, the American Civil Liberties Union, and the Prisoners Union. Support this broad is rarely found in criminal justice policy debates, where the clash of ideologies is usually a prominent feature.

Yet a coalition almost as nonsectarian as that which now undermined support for rehabilitation had in earlier years supported it. When police chiefs and prisoners are united in support of policy, it is reasonable to assume that they do not share a common picture of the likely consequences of realizing the object of their joint support. Indeed, support this broad is a pretty reliable indication that some of the participants in the policy debate do not have a clear understanding of the practical implications of their stance.

The third significant feature of the attack on the rehabilitative ideal was the almost exclusively negative character of the critique that succeeded in discrediting and overthrowing it. No effort was made to supplant the rehabilitative paradigm with a superior model or a better purpose for the use of imprisonment. The central goal of the critics was the undoing of the old order, not the sponsorship or creation of any particular new arrangement. One of the reasons the coalition opposed to rehabilitation was so diverse was that participation required *only* hostility to the prevailing standard rather than support for any specific alternative (Tonry and Zimring, 1983).

The almost exclusively negative agenda of the critics of rehabilitative policies and practices in the 1970s explains to a large extent the events following the remarkably effective attack on rehabilitation. Given that imprisonment requires some social sense of positive justification, the destruction of the dominant paradigm had to create a problem at the core of the prison enterprise. The successful attack on rehabilitation did in fact create the need for a new consensus about the purposes of imprisonment. However, there was little agreement on the positive aims and proper purposes of imprisonment among the critics of rehabilitation.

III.   Ideology and Incapacitation in the 1970s

Liberals in crime control policy tend to dislike prisons and, as a consequence, to distrust claims regarding their positive functions advanced to justify their use. Crime control conservatives, by contrast, see prison sentences as morally appropriate and thus tend more easily to accept claims that imprisonment achieves crime prevention. In accord with this traditional split, the critics of rehabilitation were sharply divided about what the desirable or optimal size of a prison system should be in a post-rehabilitation era.

The liberal view of the appropriate scale of imprisonment was taken to its logical extreme by the National Council on Crime and Delinquency, which argued in 1973 that "in any state [in America] no more than one hundred persons would have to be confined in a single maximum security institution" (1973:456). Conservatives, on the other hand, favored an expansion in prison sentences as a consequence of the abandonment of the focus on rehabilitation as the justification for imprisonment. "Since society clearly wishes its criminal laws more effectively enforced," said James Q. Wilson, "this means rising prison populations, perhaps for a long period" (1975:173), and he maintained that "the gains from merely incapacitating convicted criminals [might] be very large" (1977:22).

Do such contrasting views of the appropriate scale of imprisonment reflect different conclusions about the proper purposes to be served by imprisonment? We think not. Both the minimalist and the expansionist rationale for imprisonment in the 1970s seem to have been grounded in the notion of crime prevention through incapacitation as the primary justification of imprisonment. In the case of liberals, prison was to be reserved for a select group of especially dangerous repeat offenders in regard to whom social defense required an incapacitation strategy. The National Council on Crime and Delinquency drew a somewhat amorphous distinction between dangerous and nondangerous offenders and maintained that only the dangerous required imprisonment. Two types of dangerous offender were defined: "(1) the offender who has committed a serious crime against a person and shows a behavior pattern of persistent assaultiveness based on serious mental disturbances and (2) the offender deeply involved in organized crime" (National Council on Crime and Delinquency, 1973:456). The National Council on Crime and Delinquency emphasis on the especially dangerous offender paralleled the earlier focus of so-called habitual-offender statutes on persistent recidivists. But the important difference was that the National Council on Crime and Delinquency program wanted only this group in prison while the habitual-offender statutes assumed many different kinds of criminals would be imprisoned for other reasons.

The conservative view of the function of imprisonment in a postrehabilitation age took the form of a strategy of general incapacitation that would achieve large aggregate crime prevention gains by imprisoning substantial numbers of run-of-the-mill felons. Viewed from this perspective, some early claims for incapacitation were greeted with considerable enthusiasm two years prior to their publication. James Q. Wilson wrote, regarding one such claim—the Shinnar estimate—that "even assuming they are overly optimistic by a factor of two, a sizable reduction in crime would still ensue" (Wilson, 1973:9; Shinnar and Shinnar, 1975).

It would be easy to exaggerate the degree to which liberal and conservative critics of rehabilitation subscribed to incapacitative strategies for two reasons. First, not all of the liberal critics of the rehabilitative rationale for imprisonment embraced incapacitative premises (e.g., Morris, 1974; Messinger 1982). Second, the "special dangerousness" criteria did not appear to be strongly adhered to by proponents of the National Council on Crime and Delinquency, but seem rather to have been endorsed as a means to achieving reduction of the prison population.

The authors of the National Council on Crime and Delinquency proposal gave only loosely defined criteria for imprisonment and made no mention of appropriate terms of imprisonment or conditions of release. Nor did the proposal refer to any data to justify the very small numbers nominated for imprisonment. It is clear that for the authors of the policy statement the smallness of the number of persons who "would have to be confined" (National Council on Crime and Delinquency, 1973:456) was far more important than either the criteria for imprisonment or the categories of offenders eligible for imprisonment.

Nevertheless, because participants on both sides of the ideological debate on crime control accepted some form of incapacitation as a residual rationale for imprisonment, it was unlikely that either ideological camp would place a high priority on careful scrutiny of either the ethical requirements or the empirical aspects of incapacitative imprisonment. Indeed, in the argument over special or general incapacitation, the expansionist notion of general incapacitation was an easy winner. This was, in large part, the product of a public mood that was, throughout the 1980s, fearful and punitive. Once the legitimacy of incarcerating offenders for incapacitative reasons was accepted, public sentiment tended to favor strongly the expansionist version of that strategy.

However, though an expansionist victory may have been inevitable, two problems associated with the specification of dangerousness as a justification for special rather than general incapacitation contributed to the rout. First, the liberals never provided a persuasive distinction between the types of offenders suitable for incapacitation and the general run of convicted felons. The legitimacy of incapacitation once conceded, it was very difficult to come up with clear, objective criteria that would only permit its use in a highly restrictive fashion. The National Council on Crime and Delinquency did not even attempt to provide a definite formula but merely asserted that "only a small percentage of offenders in penal institutions today" really required incarceration (1973:449).

A second difficulty associated with trying to restrict the scope of incapacitation to the specially dangerous offender was that the process would require not only some sort of differential prediction of dangerousness but also the treatment in grossly different ways of individuals convicted of the same or similar offenses—two practices that had been lately discredited in the attack on the rehabilitative ideal. The kinds of procedure required to identify, discriminate between, and differentially treat the specially dangerous offender turn out to be closely analogous to the individualized treatment and punitive discretion involved in the administration of rehabilitative prison regimes. Having at least in some instances accepted the legitimacy of incapacitation as a basis for imprisonment policy, the liberal prison reductionists could provide no convincing limiting principle to serve as a barrier to expansionist domination.

IV. The Pattern of Academic Attention

A study of the patterns of academic attention to issues in criminology shows both the unchallenged salience of rehabilitation as the principal topic over time and the relative paucity of academic publications relating to incapacitation even after the rehabilitative ideal had been generally rejected. Table 1.1 reports a computerized search of five key terms related to various purposes of punishment (*rehabilitation/recidivism, deterrence,* and *incapacitation/preventive detention*) to see how often those terms occur in the titles of articles and books published between 1963 and 1989. The two social science services used were Sociological Abstracts (SOCABS), which covers the entire period, and the more complete Social Scisearch (SOSCISCH) system, which covers the period from 1972 onward.

With respect to the terms *incapacitation* and *preventive detention,* SOCABS reports no mentions during the 1960s and eleven mentions each for the decade of the 1970s and the decade of the 1980s. This is consistent with Jacqueline Cohen's 1978 review of the literature, which

Table 1.1. Frequency of Mention of Topics in Book and Article Titles, 1963–89

	1960s	1970s	1980s
Sociological Abstracts			
Rehabilitation/Recidivism	84	134	212
Deterrence/General Deterrence	12	52	94
Incapacitation/Preventive Detention	0	11	11
	1972–1979		1980–1989
Social Scisearch			
Rehabilitation/Recidivism	2,442		4,199
Deterrence/General Deterrence	213		610
Incapacitation/Preventive Detention	20		45

reported no publication prior to 1972 on these topics (Cohen, 1978: 187–243). By contrast, the term *deterrence* occurs in twelve titles in the 1960s, fifty-two titles in the 1970s, and ninety-four titles in the 1980s.

In the larger SOSCISCH database *incapacitation* is mentioned twenty times in the 1970s and forty-five times in the 1980s. In both decades, key words relating to *incapacitation* are on the short end of what is an almost logarithmic distribution, with mentions of *rehabilitation* and *recidivism* ten times as frequent as titles relating to *deterrence*, and with titles concerning *deterrence* occurring ten times as often as mentions of either *preventive detention* or *incapacitation*. In the 1980s, when belief in the purpose of rehabilitation in the criminal justice system was at a low ebb, almost one hundred titles relating to *rehabilitation* or *recidivism* occurred for every article that mentioned *incapacitation* or *preventive detention* in its title.

Even as incapacitation was becoming the dominant justification for imprisonment in the 1980s, academic attention to the topic remained small on an absolute basis, and in terms of relative number, both SOCABS and SOSCISCH have *deterrence* not only receiving more attention in the academic literature at any given point in time but also growing more substantially than *incapacitation* as a focus for analysis. Having begun the period as a stepchild without a supporting academic constituency, *incapacitation* has remained so until the present time.

Even the crude methodology that characterizes computerized title searches of journal articles points clearly to two conclusions. First, rehabilitation and recidivism were the most favored topics of academic analysis even in the aftermath of the decline of the rehabilitation ideal. Both of the social science research services report rehabilitation and individual responses as by far the principal focus of academic attention.

Second, the pattern of inattention to questions relating to incapacitation on the part of academics is a curious historical phenomenon. The number of titles dealing with incapacitation and preventive detention in

SOCABS went from zero in the 1960s, to eleven in the 1970s, and remained at eleven throughout the 1980s as incapacitation began to govern imprisonment policy; on the other hand, the number of titles relating to rehabilitation and deterrence both increased extensively over the same period. In the broad social science search system, incapacitation and preventive detention received less than 1 percent of the total volume of mentions of rehabilitation and deterrence between 1972 and 1979. From 1980 to 1989, though the number of mentions of incapacitation and preventive detention expanded from twenty to forty-five, the *proportionate* share of mentions those topics received remained at less than 1 percent.

V. The Appeal of Incapacitation

Why did incapacitation succeed to centrality as a justification for prisons? Why not general deterrence, a phenomenon that received more academic attention than restraint during the 1970s and 1980s? Why not elevate some conception of desert that justifies prison simply because criminals are bad persons who deserve serious punishment? What accounts for the rhetorical appeal of incapacitation in the years just after the eclipse of rehabilitation?

Three aspects of incapacitation help to explain the policy's special appeal in a post-rehabilitative era. First, there is a specific mechanical fit between what prisons do and an incapacitative rationale for them. Second, restraint from future crime operates as a concrete justification for imprisonment in individual cases. Third, incapacitative imprisonment seems morally appropriate because it singles out blameworthy persons for disadvantageous treatment. Each of these rhetorical advantages deserves further attention.

The need to deter could justify imprisonment or a variety of other unpleasant experiences that potential criminal offenders might wish to avoid, including fines, corporal punishment, loss of privileges, exile, and transportation. In no special sense is imprisonment uniquely appropriate as a deterrent. Similarly, the concept of desert might seem to justify a whole range of socially disadvantageous treatment, including imprisonment. However, because it is decided that some punishment is required because it is deserved, that alone does not indicate imprisonment is necessary, if other kinds of punishment of equivalent severity are available. By contrast, what imprisonment involves—locking offenders up and away from community settings—is ideally suited to the incapacitative function. Indeed, once there are serious doubts about prisons as instruments of rehabilitation, incapacitation is the only distinctive function that seems to flow from the structure of imprisonment. When banishment and capital punishment are no longer widely available, prison performs the incapacitative task better than any available alternative criminal punishment. So this nice fit between what prisons are specially designed to do and the incapacitative function, makes incapacitation a particularly plausible justification for imprisonment.

The second rhetorical advantage of imprisoning to incapacitate is that restraint seems to operate as a concrete justification for the use of imprisonment at the level of the individual case. An instructive distinction can be drawn here between incapacitation and general deterrence. Punishments are needed to deter potential criminals. But how many and what kinds of punishments? And can we ever say that it is necessary to punish John Smith rather than others convicted of similar crimes if effective general deterrence is to be achieved? By contrast, once it is granted that John Smith may offend again if he is not restrained, the need for incapacitation appears to justify not only a general regime of incapacitation but also the imprisonment of this particular miscreant. In this sense, both desert and incapacitation have a substantial rhetorical advantage over the nonspecific claims of deterrence, at least when large numbers of candidates for punishment are available.

A third advantage possessed by incapacitation as a justification for imprisonment is that it will commonly appear morally fitting and appropriate because it involves punishing particular individuals to prevent *them* from committing further crimes. In the case of general deterrence, the reason that certain individuals are imprisoned, or that prison terms are lengthened, is to prevent the commission of crimes *by other people*. The apparent advantage of the incapacitative justification is that represents the pursuit of crime prevention by locking up the right people.

It is interesting to note the degree to which the rhetoric of incapacitation is suited to the particular period of its ascendancy. The case for incapacitation at the individual level and the intuitive sense that the "right people" are being imprisoned rest on the premise that the individual who has offended once will offend again unless restrained. The implicit assumption that criminal offenders are intractable and insusceptible to change serves to justify imprisonment for the purpose of restraint on both moral and practical grounds. Indeed an image of the criminal offender as intractable was very much in fashion in the United States by the 1990s. Thus, the attack on rehabilitation encouraged a view of criminal offenders that made incapacitation appear to be a singularly suitable policy goal for prisons.

VI. Incapacitation and Prison Expansion

The rhetorical advantages of incapacitation were destined to play an important role in debates about the expansion of prison capacity in the United States in the 1990s. Inevitably, appeals to build more new prisons or to expand the capacity of existing correctional facilities have been framed in terms of incapacitation rather than deterrence, desert, or other functions of imprisonment. "The choice is clear," said the Attorney General of the United States in 1992. "More prison space or more crime" (Barr, 1992:345).

As an argument for prison construction, the specificity and palpability of incapacitation are signal advantages. Each additional prisoner represents more crime prevented as long as it is assumed that offenders will persist in crime unless confined. Thus, each new prison bed promises crime prevention in a much more concrete fashion than general deterrence. Once persistence in criminal activity is assumed, greater amounts of imprisonment *must* prevent crime and the only question at issue is the extent of that preventive effect.

For this reason, incapacitation has become both a powerful argument for the expansion of prison capacity and an almost irresistible political argument against any kind of decarcerative proposal. Just as locking up more offenders *must* reduce criminal activity by some amount, releasing large numbers of offenders or allowing them to remain outside prison *must* produce some increase in the number of crimes experienced in the community that receives them. Support of decarceration is thus the moral equivalent of approving higher crime rates, entailing a high risk of political extinction for anyone sufficiently naive to endorse the policy.

The power of this logic as a barrier to decarceration is nowhere more apparent than in the 1990 report of the California Blue Ribbon Commission on Inmate Population Management. Faced with an unprecedented increase in prison population, the Blue Ribbon Commission was created by statute in 1987 (Senate Bill 279, Chapter 1255, Statutes of 1987), its task described in the official Mission Statement as being "to determine viable strategies to deal with problems of prison overcrowding *without reducing public safety*" (Blue Ribbon Commission, 1990, Appendix B.1; emphasis added). The authorizing statute creating the commission stated categorically: "It is the intent of the Legislature that *public safety shall be the overriding concern* in examining methods of . . . heading off runaway inmate population levels," and further: "*Public safety shall be the primary consideration* on all conclusions and recommendations" (Blue Ribbon Commission, 1990:1, 9; emphasis added).

The operational problem involved in responding to a prison crowding crisis without creating any additional risk of crime in the community is that any reduction in prison population inevitably runs *some* risk of crime in the community from the released or the unimprisoned. Thus, a policy proposal from the commission would inevitably violate one of the terms of its mandate either by failing to reduce the number of prisoners or exposing the community to crime (Zimring and Hawkins, 1992:17–28).

The role of incapacitation in the politics of prison expansion deserves emphasis because the critical examination of incapacitation, a process finessed by its sudden emergence as the dominant justification for imprisonment, inevitably enters the debate about prison construction. Unchallenged claims for incapacitation justify virtually unlimited expansion of imprisonment in a country where the majority of convictions after felony arrest still lead to nonimprisonment outcomes. No obvious point of equilibrium in prison construction can be reached once the hallmark for criminal justice policy becomes zero risk to public safety. So the story told in this chapter is that of a debate postponed rather than avoided. The political claims made on behalf of incapacitation are too radical in their operational consequences to remain unchallenged.

2

The Short History of an Idea

When President Gerald Ford addressed Yale University Law School's Sesquicentennial Convocation Dinner in April 1975, he devoted most of his speech to the crime problem. "The core" of the problem, he said, consisted of "relatively few persistent criminals . . . a very small percentage of the whole population." The solution to the problem, he suggested, was "to get them off the street."

> The crime rate will go down if persons who habitually commit most of the predatory crimes are kept in prison for a reasonable period . . . because they will not then be free to commit more crimes . . . one obvious effect of prison is to separate law breakers from the law-abiding society. (Ford, 1975:591–93)

The president's words might have been regarded, and by some members of his audience no doubt were, as representing a simplistic approach to a complex problem. Removal from a community or country simply as a preventive measure is, of course, society's most primitive form of social defense, exemplified historically in the death penalty and extraterritorial transportation. On the other hand, it is possible to regard his remarks as a remarkable example of presidential prescience and sophistication. Reviewed in the context of academic and official discussion of penal policy in 1975, the concept of incapacitation—that is, removing an offender from society and thereby physically preventing that offender from committing crimes in society—as the primary justification for imprisonment was about to become a significant theme in the discussion of penal policy and correctional strategy.

When President Ford delivered his Yale address, very little had been written and much less remembered about incapacitation as a motive and effect of penal policy. In this chapter, we shall review the short history of incapacitation in two installments. In section I, we discuss three eras of discourse on incapacitation: (1) the analysis by Jeremy Bentham at the beginning of the nineteenth century; (2) the later emphasis on incapacitation in dealing with habitual offenders; and (3) the discussion of incapacitation in "modern" literature of the 1970s and 1980s.

Section II addresses the post-1975 discussion of incapacitation as a literature. Whereas later chapters of this book will discuss the contribution of various authors to the discussion of research methodology and theories of incapacitation effects, this chapter is concerned with the development of an incapacitation literature as an historical process. We identify cyclical patterns in the published discussion of incapacitation and show how these patterns can be used to anticipate future development in scholarly and policy emphasis.

Significantly, before 1991 the published writing on this topic in the English language is sufficiently sparse to be covered in a single chapter, and most of that writing is concentrated in the years after 1972. The protracted inattention to the topic previously is, in retrospect, more remarkable than the attention it has since received.

I. Bentham and Beyond

The intellectual history of incapacitation is astonishingly short. At the beginning of the nineteenth century, Jeremy Bentham said all that was to be said on incapacitation for a hundred years in his analysis of the competition between the penitentiary and transportation.

Bentham

Jeremy Bentham's detailed analysis of incapacitation is of importance to modern scholars not least because they must address why matters he identified were ignored for so long and suddenly become so important in later debate.

In Bentham's *Panopticon versus New South Wales* (subtitled *The Panopticon Penitentiary System and the Penal Colonization System Compared*), written in 1802, he dealt at considerable length with the subject of "incapacitation for fresh offenses." As a critic of transportation and admirer of the American penitentiary system, he recognized that for supporters of transportation, its incapacitative effect was a major attraction. He therefore felt it necessary to argue that imprisonment was no less effective than transportation in achieving that effect, for he had included incapacitation in his statement of the principal "objects and ends of penal justice . . . the ends to which the [penal] measures adopted ought to tend in a direct course."

That statement incidentally is characteristically dogmatic. Said to be derived "from Blackstone and from everybody" and described simply as "belonging to the ABC of legislation," it is not advanced as a subject for discussion but rather as self-evidently correct. Bentham gives as the first object of penal justice: "prevention of similar offenses on the part of individuals at large, viz. by the repulsive influence exercised on the minds of bystanders by the apprehension of similar suffering in case of similar delinquency"—in other words, general deterrence. The second object is reformation, which is defined as "prevention of similar offenses on the part of the *particular individual* punished in each instance, viz. by curing him of the *will* to do the like in future."

The third object is incapacitation: "prevention of similar offenses on the part of the same individual by depriving him of the *power* to do the like" (Bentham, 1843:174; emphasis in original). His lengthy argument in support of imprisonment as more effectively incapacitative than transportation covers sixteen pages in double columns of extremely small print in the standard edition of Bentham's works. The argument begins with some rather ponderously ironic passages.

> III. Third object or end in view—*Incapacitation,* rendering a man incapable of committing offenses of the description in question any more: understand in the present instance *in the same place*—the only place (it should seem) that was considered as worth caring about in this view.
>
> In this object was seated, to all appearances, the strong hold and main dependence of the system: of reformation it would (I dare believe) have been acknowledged in a whisper there was nothing meant but the form: it was a mere make-believe. In the expedient employed for rendering it impossible for a man to do any more such mischief in the only spot in the world worth thinking about, consisted the sum and substance of the new system of compulsive colonization.
>
> This contrivance was as firmly laid in school-logic as could be wished. Mischievously or otherwise, for *a body to act in a place,* it must be there. Keep a man in New South Wales, or anywhere else out of Britain, for a given time: he will neither pick a pocket, nor break into a house, nor present a pistol to a passenger, on any spot of British ground within that time. (Bentham, 1843:183; emphasis in original)

"On these principles," Bentham goes on to say, "how the people sent there behaved while there, was a point which, so long as they did but stay *there,* or, at any rate did not come back *here,* was not worth thinking about" (Bentham, 1843:183; emphasis in original).

However, he points out that of the length of sentences to transportation "beyond comparison the most common [is] seven years" and the "number of returnees, whatever it may have been hitherto, it may naturally be expected to be greater and greater the longer the establishment continues" (1843:183, 192). Moreover, he argues that although the indigent, unenterprising, and "least dangerous species of malefactor" would stay in New South Wales, "for want of being able to employ with success those means of escape, which his more ingenious, or more audacious, and on either account more dangerous comrades, make such abundant and successful use of," those most likely to return would be "precisely those, from whom, the most mischief is to be apprehended" (1843:193–94). It is relevant to mention in this connection that much earlier, in 1725, Bernard Mandeville, criticizing various aspects of the transportation system, complained that convicts escaped before shipment or returned to England prematurely (Mandeville, 1964:47–48).

Bentham concludes, not altogether persuasively perhaps, that

> *during* the continuance of the penal term, at any rate, the advantage, so far as the article of *incapacitation* for fresh offenses is concerned, may, I flatter myself, be stated as being clearly enough on the side of the penitentiary establishment. Even in an ordinary prison, an escape is not a very *common* incident; under the new and still more powerful securities of so many sorts, superadded to the common ones, in a prison upon the panopticon plan, I have ventured to state it as, morally speaking, an *impossible* one. (1843:194; emphasis in original)

Yet it is clear that Bentham did not regard incapacitation as of primary importance. Of all the objects of punishment by imprisonment,

> example [i.e., general deterrence] is beyond comparison the most important. In the case of *reformation* and *incapacitation*, for further mischief, the parties in question are no more than the comparatively small number, who having actually offended, have moreover actually suffered for the offense. In the case of *example*, the parties are as many individuals as are exposed to the temptation of offending . . . in others words, all mankind.(1843:174; emphasis in original)

Although the primacy of deterrence was to be vigorously challenged before the end of the nineteenth century, the subordinate position of incapacitation was to be accepted almost without question for more than a century and a half.

Thus, Lionel Fox in his examination of the principles of imprisonment as a legal punishment, published 150 years later in 1952, dismisses incapacitation, which he refers to as "Prevention," as "*a mere abstraction*" because "in the present state of penal law and public conscience sentences are not in practice—excepting always the special case of the persistent offender—based on this conception" (Fox, 1952:17; emphasis added). Moreover, even today incapacitation sometimes escapes notice. It is, for instance, not even mentioned in a recent account of "the classic objectives of punishment" (Sharpe, 1990:6–15).

In both England and America, the great debates about penal policy in both the nineteenth and twentieth century were principally about whether retribution and deterrence or reform should be the major principles under which the treatment of prisoners should be regulated. At the quinquennial International Prison Congresses organized by the International Penal and Penitentiary Commission, beginning in 1872 and continuing until 1950, when there was discussion of penological ideas, it commonly focused on prioritizing these general principles (Ruggles-Brise, 1924). As late as 1972, it was generally assumed "that the primary conflict in prison is between reformation and deterrence" (Thomas, 1972:2). For almost two centuries after its mechanisms and limits had been identified, incapacitation played no role in the debates about justification for imprisonment. Of course, when the subject did become important, Jeremy Bentham's words had been long forgotten.

The Exception That Proves the Rule: Restraint and the Habitual Offender

Although the incapacitation effect was being ignored as a general justification of imprisonment, it was a significant feature of policy directed toward one group called habitual or career offenders, distinct from the general run of offenders. The perspective first became prominent in legislation after the turn of the twentieth century and recurred in policy proposals in the 1970s. In each case, the habitual-offender initiatives confirmed the deliberate lack of emphasis on incapacitation as a general policy right up to the 1970s.

In the early decades of the twentieth century and in regard to a limited category of offender variously defined as dangerous, persistent, habitual, or professional, the principle of incapacitation served to provide a rationale for preventive detention, viewed as a measure of public security rather than traditional legal punishment. According to Max Grünhut,

> "incapacitation" of dangerous persistent criminals by long-term, if not lifelong, segregation was the outcome of the science of criminology. Lombroso's doctrine of the "inborn criminal" allowed no alternative to permanent isolation, once congenital dangerousness had been diagnosed. Likewise von Liszt, of the crimino-sociological school, regarded "the vigorous struggle against habitual criminality [as] one of the most urgent tasks for the present." The incorrigible criminal therefore "should be incapacitated from crime by imprisonment for life or for an indeterminate period." (1948: 386)

Whether or not the influence of "the science of criminology" was as great as Grünhut thought, he goes on to demonstrate the way in which preventive detention for dangerous or habitual criminals came to be accepted and embodied in legislation throughout Europe and America. Beginning with the English Prevention of Crime Act of 1908, the criminal codes of Czechoslovakia (1926), Sweden (1927), Norway and Yugoslavia (1929), Belgium, Italy, and Denmark (1930), France and Poland (1932), Germany (1933), and Switzerland (1937), traditional legal punishment was, for persistent offenders, either replaced or supplemented by preventive measures imposed for public security reasons (Grünhut, 1948:392).

In America preventive detention was achieved by the prolongation of prison sentences for specially designated classes of offenders. Between 1920 and 1930, twenty-three states enacted laws increasing the severity of sentences for recidivists. Thus, the Penal Code of California was amended in 1927 to provide that a prisoner convicted after two previous convictions for certain specified crimes (e.g., robbery, burglary, arson) incurred imprisonment for life as an habitual offender with no possibility of parole before the expiration of twelve years, and that any fourth conviction for a felony incurred imprisonment for life with no prospect of parole. In the same spirit, the Baumes Law added to the Penal Law of New York in 1926 provided that a prisoner convicted of a felony for the fourth time should incur imprisonment "for the term of his natural life" (Grünhut, 1948:394). By 1979 recidivist statutes were in force in forty-four states, the District of Columbia, Puerto Rico, the Virgin Islands, and at the federal level (American Criminal Law Review, 1979:275).

The application of the incapacitation principle in sentences of preventive detention was everywhere limited to recidivists or habitual criminals. Although neither of these labels has any generally accepted definition, they have in this punitive context commonly been applied to those who have repeatedly committed criminal offenses of a serious nature. Lionel Fox asserts that "there must be something more than mere repetition of offense: It is generally accepted that repetition must be indicative of a significant character trait" (Fox, 1952:298). And Max Grünhut claims that in the case of "*habitual criminals proper,*" the offender's crimes "are rooted in his own personality, attitude, and inclination" (1948: 389–90; emphasis added). In neither case do the authors explain either the necessity for the possession of a particular character trait or personality type in addition to a serious criminal record in order to be regarded as an habitual criminal, or how such possession could be verified. The legislative categories, moreover, usually made repetition of offenses a sufficient condition for special punishment.

What is clear, however, is that the habitual criminal category for which incapacitative or preventive measures were seen as necessary was intended to be a restricted one. "The majority of legal systems," says Grünhut,

> combine two elements. They require recidivism of a certain intensity and seriousness as a legal condition, without which preventive detention is not to be considered. Within these limits, they leave it to the discretion of the court to decide whether the recidivist in the light of his antecedents, character, previous prison and parole experience is an habitual criminal so dangerous that the protection of society calls for long-term segregation. (1948:395)

The Twelfth International Penal and Penitentiary Congress held at The Hague in 1950 passed a resolution, the first paragraph of which says: "Traditional punishments are not sufficient to fight effectively against habitual criminality. It is therefore necessary to employ other and more appropriate measures" (Fox, 1952:440). In practice, however, the material difference between the traditional punishment of imprisonment and the "more appropriate measures" was slight (except that the latter were frequently of indeterminate duration). Indeed, Grünhut observed, "No existing system has succeeded in differentiating between ordinary prison routine and the regime applicable to preventive detention" (1948:399). What was different, in respect of this variously and imprecisely defined class of offender, was the rationale for confinement.

The emphasis on incapacitation for habitual offenders can be viewed as not inconsistent with the dominant role of rehabilitation for other offenders. Reserving "habitual" status to those who had been often convicted and punished made the incapacitation sought for the group a last resort for those for whom penal measures aimed at rehabilitation and threats intended to deter, had repeatedly failed. Thus purposes of punishment with higher priority had already been tried, and their higher priority had been respected, when restraint was adopted as a fallback.

For the great mass of prisoners both in the penological theory and penal practice, the treatment of offenders was based on either deterrent or reformative principles. In the early part of the twentieth century, these were seen by some to be quite incompatible. At the 1910 International Penal and Penitentiary Congress held in Washington, D.C., there was sharp disagreement between the English and American delegations; with the Americans being accused of having "swung too violently away from the classical traditions of punishment" by giving reformation priority over deterrence (Leslie, 1938:163). And in 1947, James V. Bennet, Director of the U.S. Federal Bureau of Prisons, defined "the basic problem" as "how to use imprisonment, which is inherently a symbol of punishment, to achieve a purpose almost completely antithetical to punishment—rehabilitation" (U.S. Department of Justice, Bureau of Prisons, 1947:4).

The set of initiatives to imprison a small number of "career criminals" that was proposed in the mid-1970s provides a striking if unacknowledged parallel to the habitual-offender statutes. In the case of career criminals, a small number of offenders were supposed to represent a special threat requiring sustained isolation in a prison for incapacitation. In the mid-1970s, this number was estimated at 10,000, under 5 percent of the then-current nationwide prison population.

This later emphasis on career criminals—the programmatic context for Gerald Ford's remarks cited at the beginning of the chapter—was typical of habitual-offender approaches in two respects. First, the mid-1970s program did not refer to any earlier efforts to isolate dangerous offenders. The statements in support of the Ford administration proposal ignored the experiences of habitual-offender programs throughout the developed world. This is typical of the entire dialogue on incapacitation. It is as if proposals to use prisons for special or general programs of restraint existed in a contextual void, their authors almost never providing historical reference or perspective.

Second, programs of incapacitation for special groups reject by implication the notion that incapacitation should be the dominant purpose of imprisonment for other offenders. In this sense habitual-offender laws could be said to be consistent with an emphasis on rehabilitation for all other offenders. Perhaps also this emphasis might explain why the earlier efforts to incapacitate special groups were not generalized, as happened in the mid-1970s.

The Modern Emphasis

It was not until the 1970s that Jeremy Bentham's nomination for third place in the list of objects and ends of punishment came to be not only seriously considered for promotion, but nominated for primacy. In 1972, J. E. Thomas advanced the thesis that the primary task of the prison system in respect of *all* prisoners was not deterrence or reform, but containment or control:

> For several hundred years English society has agreed that people who reject certain social values must be removed from that society. Outlawry in the Middle Ages, transportation from the seventeenth century, and imprisonment from the nineteenth are all expressions of that agreement. Imprisonment, other than as a preliminary to transportation or death, is a relatively new method of dealing with offenders. Society has defined the need for the removal of the criminal, and the prison system, as an organization, has come into being to achieve that task. The court warrants instruct the prison authorities to hold the prisoner. These documents say nothing about treatment. (Thomas, 1972:5)

Thomas's argument, which he claims to be "substantially relevant to prison systems of all advanced societies" (1972:xiv), is an odd mixture of both apriorism and empiricism. It is both deductive and inductive. Thomas begins with the concept of the primary task defined by A. K. Rice in his work on social organization as the "task which it [the organization] is created to perform" (Rice, 1958:32). "Clearly," says Thomas, "the attempt to define the primary task is an essential preliminary to any investigation of organizational structures." He acknowledges that in some organizations there may be more than one goal and also conflict about priorities; "but there can only be one primary task" (Thomas, 1972:4).

The definition of the primary task is arrived at by asking, What constitutes failure? How is it recognized that the organization is unsuccessful? To answer those questions, Thomas borrows from Peter Nokes's work on professionalism in welfare work the concept of "the manifest disaster criterion" (Nokes, 1967:19–22). Does the prison system, Thomas asks rhetorically, "have a manifest disaster criterion, [the]

definition of which could lead to a definition of a primary task?"
Acknowledging that a prison system may have conflicting goals, he
argues that if the principle of demonstrable failure is applied to it, then
the primary nature of one goal becomes evident. "[S]ociety has called
the prison system into being to ensure removal of recalcitrant members
of society." How then is success or failure measured? The answer is clear:

> The manifest disaster criterion must be failure at successful contain-
> ment and control of prisoners . . . loss of control must be avoided. The
> signs of loss of control are escape or riot, both of which are easily visi-
> ble to society. They are the manifest disaster criteria which demon-
> strate task failure. (Thomas, 1972:5,7)

It is clear that there is an element of logical legerdemain involved
in Thomas's argument, but what is significant is the nomination of control or containment as the prison system's primary task. Prison history,
says Thomas, has been presented as a battle between those who wished
for reform and those who opposed it on the grounds that deterrence
should be the object of prison organization. Although this debate had
occupied the forefront of writing about prisons, "the debate about the
treatment of prisoners is a debate about the means of achieving a *secondary* goal." That "*secondary* task" for the prison system is defined by
Thomas thus: "that the prisoner, having undergone the experience of
imprisonment, should not return to prison" (Thomas, 1972:6).

Another book, which appeared at about the same time but which
was far less schematic than Thomas's, was *Punishment, Prison, and the
Public* by the Vinerian Professor of English Law in the University of
Oxford, Rupert Cross. Subtitled *An Assessment of Penal Reform in Twentieth-Century England by an Armchair Penologist*, it conveyed a skeptical view
of the traditional justifications of punishment in general and of imprisonment in particular.

Cross allowed that "retribution and general deterrence" were "*com-
prehensible* justifications of punishment, however wrong-headed some of
us may consider some forms of the former to be, and however skeptical
others may be about the latter" (Cross, 1971:46; emphasis added).
Unlike Thomas, Cross did not think incapacitation was the primary purpose of imprisonment. However, he was very skeptical about the possibility of reform in prison and argued that one of the principal justifications for sending people to prison was "for the protection of the public
from their depredations while they are in prison."

"If analogies have to be drawn," he said, "prisons are more like cold
storage depots than either therapeutic communities or training institutions" and "the temporary incapacitation of the criminal" is, he said, an
important "element in every [prison] sentence." He added, however, that

> unless we are prepared to tolerate the incarceration of a certain num-
> ber of people for the whole of their natural lives, the protection of the

public from the possible future depredations of the offender by means
of a prison sentence must always be a matter of degree. (Cross,
1971:85–86, 159)

Thus, one hundred seventy years after the publication of Bentham's
Panopticon versus New South Wales, renewed attention was directed to the
isolating and incapacitative function of imprisonment. The nature of
that function was well put by Leslie Wilkins in an address to the American Philosophical Society in 1974:

> It is not unreasonable for persons who have suffered from some crime
> to demand that the offender "get out of here." When "out of here" did
> not mean into some similar society (such as the next state), and trans-
> port was slow, there were several variations of the general theme of
> "out of here." Few areas of the world can today employ such methods.
> The prison has to suffice. (1974:246).

The revival of interest in incapacitation in the 1970s appears to
have been both ahistoric and disjunct. Thomas makes no reference to
Bentham's writing on this topic and neither Bentham's nor Thomas's
work is mentioned in or listed in the references to Jacqueline Cohen's
review of the literature on the incapacitative effect of imprisonment.
Nor is either mentioned in any of the articles and books cited by
Jacqueline Cohen. The first item in her list to deal chronologically with
incapacitation—a discussion paper by Jeffrey Marsh and Max Singer
suggesting that expanding the use of imprisonment might result in substantial gains in reduced crime—apparently had no precursor in America (Cohen, 1978:187–243; Marsh and Singer, 1972). Since then, however, the incapacitative effect of imprisonment has been the subject of
considerable attention.

The exposition of modern prison incapacitation theory is frequently couched in rebarbative language embellished with quasi-mathe-
matical symbols and notation schemes that tend to obscure rather than
clarify meaning. Yet the basic idea is very simple: for so long as offenders are incarcerated, they clearly cannot commit offenses outside
prison. Thus, any imprisonment policy will have an incapacitative effect
as long as any of those imprisoned would have been offending if they
were at liberty. "The fundamental notion of incapacitation" is, as Alfred
Blumstein puts it, "taking a slice out of an individual criminal career"
(1983:874).

The two earliest publications in Jacqueline Cohen's review of the literature dealing with incapacitation reflect opposing views about the
potential gains in increased incapacitation if imprisonment were to be
increased. In "Soft Statistics and Hard Questions," which appeared in
1972, Jeffrey Marsh and Max Singer argue that the potential gains in
reduced crime from increased use of imprisonment might be substantial. By contrast, Stevens Clarke's "Getting 'Em Out of Circulation: Does

Incarceration of Juvenile Offenders Reduce Crime?," published in 1974, offers a low estimate of the incapacitative effect of imprisonment.

A significant difference between these two studies is that whereas Marsh and Singer's estimates are based on the assumption that offenders vary in their offending rates and that high-crime-rate offenders are more likely to be incapacitated, Clarke, uses a simpler model, which assumes average crime rates and the application of a uniform sentencing policy. He allows for different age-specific arrest and incarceration rates, but they are assumed to apply uniformly within each age category.

Marsh and Singer consider a hypothetical distribution of individual crime rates among criminals. Examining robberies in New York, the researchers partition the population of robbers into six groups, with assumed rates ranging from 1 robbery per year to 250 per year. Robbers with higher offense rates are assumed more likely to be detected and thus imprisoned. Taking their expected number of robberies in a year after their release as an index, the researchers calculate that the possible reduction in robberies as a consequence of imprisoning these robbers for an extra year would be in the range from 35 to 48 percent of the total robberies in a year (Marsh and Singer, 1972:10–11).

Stevens Clarke uses data from records for a cohort of 9,945 Philadelphia boys to estimate the proportion of FBI index crimes averted by incapacitation. Clarke estimates the average annual arrest rate from juvenile offenders while free; then applying these rates to the periods of incarceration, he estimates the number of arrests prevented by the incarceration of these juveniles. He determines the incapacitative effect of the then-prevailing use of incarceration for juveniles to be from 5 to 15 percent of reported index crimes by juveniles and from 1 to 4 percent of all (adult and juvenile) reported index crimes (Clarke, 1974:534).

Like Clarke's study, another early paper on incapacitation, "The Incapacitative Effects of Imprisonment: Some Estimates," by David Greenberg, provides low estimates of the incapacitative effect. Greenberg uses official data on arrests to estimate the incapacitative effect of then-prevailing incarceration policies. Like Clarke, he posits a degree of uniformity in his data: he assumes all criminals to have the same individual crime rate and the same probability of apprehension given a certain crime. Greenberg offers a number of different procedures for estimating the additional incapacitative benefits to be derived from increasing the amount of time served in prison. First, he calculates the number of additional index offenses that would be expected from lost incapacitation if the prison population were reduced. He estimates that if the daily average prison population were reduced from 200,000 to 100,000 and the amount of free time thereby increased by 100,000 more years, there would only be a 0.6 to 4.0 percent increase over the then-currently estimated total of 8.34 million offenses. Applying these results to possible reductions in crime because of expanded use of imprisonment, he finds that an increase of one year in the current average prison term of two years (a 50 percent increase) would *ceteris paribus* only produce a 0.6 to 4.0 percent decrease in crime from incapacitation (Greenberg, 1975:566–67).

Greenberg provides another low estimate of incapacitative effect, derived from the rate of parolee returns to prison for new index offenses. The number of index offenses committed per person can be estimated from the data on returns to prison, given the clearance rate (which converts known offenses to arrests), the nonreporting rate, and the probability of return to prison given arrest for parolees. After correcting for variations in the reporting rate and the reduction in offenses that is due to custody, Greenberg estimates that individuals commit just over two index offenses per year. This would amount to only a 1 to 2 percent increase in index offenses nationally from lost incapacitation if the prison population were reduced by 50 percent to 100,000 (Greenberg, 1975:570–71).

An article by Reuel and Shlomo Shinnar entitled "The Effects of the Criminal Justice System on the Control of Crime," published in 1975, provides, according to Cohen (1978) a "model [that] represents the best approach to estimating the incapacitative effect available to date." The article, which also made substantial claims for the incapacitation impact of increasing imprisonment, featured prominently in the policy debate on prison even prior to its publication in 1975 (see e.g., Wilson, 1973). Shinnar and Shinnar's model of the criminal career postulates that if criminals are arrested, convicted, and sent to prison during their career, their total number of crimes will be reduced by the same proportion that their free time is reduced. On this assumption, it is possible to estimate the percentage reduction in crime produced by a particular imprisonment policy.

Using data for New York state, the authors compare the relative effectiveness of the criminal justice system in 1940, 1960, and 1970. The expected prison stay per crime committed is estimated from the average daily prison population divided by the number of crimes. The authors calculate that in 1940 the effective reduction from potential crime is between 75 and 85 percent. By 1960 this is down to a 56 percent reduction; and by 1970 the incapacitative impact of imprisonment has been reduced to 20 percent (Shinnar and Shinnar, 1975:604).

The Shinnars' model paper is simple: Incapacitation is the effect achieved by incarcerating offenders and thereby preventing them from committing crimes in the community. The number of crimes committed at any time will be reduced by the crimes prevented by the imprisonment of some offenders. It follows that if more convicted criminals were sent to prison or they were sentenced to longer average terms, more crimes would be prevented. Thus stated, this part of the authors' argument for the incapacitative effect is almost irrefutable: a series of analytic or necessarily true propositions.

Although it may derive plausibility from its reported increase in crime rates and proportional decrease in incarceration rates and average sentence lengths, the model does not involve any information based on observation in the real world, or any assumptions other than that the offenders who are imprisoned would have continued to commit crimes if they had remained free. No specific image of the offender is supposed. Criminals are nonentities: ciphers who are all assumed to commit crimes at the same rate, and to be subject to arrest, conviction, and imprisonment, given conviction, with the same probability. For all criminals, if there is no imprisonment, the expected number of crimes in a criminal career is the same. If a criminal is arrested, convicted, and imprisoned, his total number of crimes will be reduced by the proportion that his free time is reduced. If an offender is incarcerated for 50 percent of his career, the number of crimes he commits will be 50 percent less than he would have committed if he had never been incarcerated.

The Shinnars' scheme is essentially deductive rather than inductive. It is based on reasoning from a number of assumptions rather than from empirical observation. The conclusion that the rate of serious crime would be only one-third of what it is if every person convicted of a serious offense were imprisoned for three years is not derived from any data relating to serious offenders or their offenses. The conclusion that violent crime rates would be reduced by as much as 80 percent if every person convicted of a violent crime were imprisoned for five years (Shinnar and Shinnar, 1975:607) is not supported by any evidence about the criminal careers of violent offenders. This is not surprising because the requisite evidence was not available.

The work of Van Dine, Dinitz, and Conrad, by contrast, represents "a focused empirical attempt to evaluate the effectiveness of the incapacitation strategy" (Van Dine et al., 1977:22). They set out "to fill an empirical gap" and to address "the basic question . . . : How many actual offenses might have been prevented by sentencing policies which are designed for the purpose of incapacitating the dangerous offender?" (Van Dine et al., 1977:24). Their study examines the prior criminal records of all adults arrested for violent felonies (murder, rape, robbery, and aggravated assault) in Franklin County, Ohio (Columbus), whose cases were disposed in court during 1973. The authors test a series of hypothetical sentencing policies from the most stringent—a five-year mandatory sentence on any prior felony conviction, whether violent or not—to far lighter sentences to determine the effects of adult incapacitation-oriented policies on this cohort of violators. The authors conclude that the most stringent option would have prevented no more than 4.0 percent of the violent crime in Franklin County in 1973.

A number of critics, however, have pointed out that in arriving at the 4.0 percent reduction, Van Dine and his associates limited the crimes potentially averted by the five-year prison terms to those reported offenses that resulted in charges in 1973, whereas their base figure of crimes included *all* reported offenses, some of which must have been committed by the arrestees (Boland, 1978; Johnson, 1978; Palmer and Salimbene, 1978). In responding to this criticism, Van Dine and his associates acknowledged its validity and said that their 4.0 percent estimate represented "a minimum value for incapacitation of the vigorous kind we postulated" (Van Dine et al., 1978:137).

In their later work, the Van Dine team adopted an alternative treatment of reported but uncleared offenses and concluded that a five-year mandatory prison term imposed on any felony convictions would have prevented 17.4 percent of the 638 violent felony arrests in 1973 (Van Dine et al., 1979:63). However, another empirical study by Petersilia and Greenwood, based on prior criminal history data for a random sample of 625 persons convicted of serious offenses in District Court between 1968 and 1970 in Denver, Colorado, produced a substantially greater estimate of crime reduction through collective incapacitation. They reported that five-year mandatory prison terms imposed after any felony conviction as an adult could have prevented 31 percent of convictions following arrests for violent offenses, although if the five-year terms were restricted to repeat felony convictions, the reduction decreased to 16 percent (Petersilia and Greenwood, 1978:608–610).

In a later review of all the available estimates of the effects of alternative collective incapacitation policies, Jacqueline Cohen concluded that the evidence suggested

> a generalized expectation of a potential reduction in serious crimes of about fifteen percent from mandatory five-year prison terms imposed after any felony conviction and about a five percent reduction from the same terms imposed only after repeat felony convictions. To the extent that offenders with prior felony convictions have a higher chance of arrest for their subsequent crimes than do other offenders (perhaps because they are known to the police and more likely to be considered as suspects), these estimates will overstate the potential incapacitative effect of these sentencing strategies on crimes. As such, they represent upper bounds on the crime reduction from incapacitation that can be expected from fairly stringent policies of mandatory five-year prison terms after conviction. (1983:27)

When the National Academy of Sciences Panel on Research on Deterrent and Incapacitative Effects reported its conclusions in 1978, it noted that

> The literature on incapacitation contains a number of studies offering widely divergent estimates of the incapacitative effect of imprisonment. Some argue that these incapacitative effects are negligible, while others claim that a major impact on crime is possible through increases in the use of imprisonment . . . The principal disagreement is over the value of the individual crime rate, an issue that can only be

resolved empirically . . . All of the available models for estimating the incapacitative effect rest on a number of important, but as yet untested, assumptions that characterize individual criminal careers. (Blumstein et al., 1978:66–67)

Five years after the Shinnars' paper was published, Joan Petersilia reported:

Little is known about the extent and types of crime committed at different stages of criminal careers. Few reliable estimates have been made of the probabilities of arrest, conviction, and incarceration. How these probabilities vary over the career, among offenders, or among offense types is not known. These questions are central to the assessment of criminal justice policies that aim to increase deterrent or incapacitative effects. If crime commission and arrest rates differ significantly among offenders and over the career, the effect of longer sentences on overall crime will depend greatly on who is incarcerated for how long.

Little is known about how and when the criminal career begins, or how long it is likely to last, why criminal careers persist, and why some persons abandon criminal careers early, others continue into adult crime, and still others begin crime careers late in life. (1980:325–26)

Studies that were relevant to some of the questions addressed by Petersilia emerged from the Rand Corporation Habitual Criminals Program. Started in 1974, the program was originally designed to determine the size and characteristics of the habitual-offender population and to examine their interactions with the criminal justice system. In response to policy concerns, the focus of the research reflected "a growing interest in incapacitation as a policy goal, and a concentration on career criminals as a specific means of crime reduction" (Petersilia, 1980:327).

The first of those studies, *Criminal Careers of Habitual Felons* by Joan Petersilia, Peter Greenwood, and Marvin Lavin (1977), which laid the groundwork for subsequent Rand surveys of much larger samples of inmates, gathered data on the criminal careers of forty-nine prison inmates in California prisons, all of whom were armed robbers serving at least a second prison term. They were selected as exemplars of serious recidivist offenders. The data were gathered in extensive personal interviews with the offenders and also from official criminal records. The self-reported crime data passed some simple tests of validity (Petersilia, 1978).

In personal interviews, inmates were asked about their frequency of offending and about their prior criminal record for nine offenses as juveniles, as young adults, and prior to the start of their current sentences. They were also asked about aspects of their personal histories, such as family circumstances, school and employment experiences, drug and alcohol use, personal motivations for crime, and styles of committing crimes.

The findings on average levels of offending over time for the forty-nine offenders are summarized in table 2.1. As one reads down the table, the offense classes become more inclusive; the total rate includes offenses in any of the nine offense types surveyed.

It is notable that except for violent offenses, the reported monthly rates declined significantly as the offenders got older. According to the authors, the slight increase in monthly rates with age for violent offenses is "undoubtedly a consequence of our sample selection criterion that the interviewees be active robbers in the adult career period" (Petersilia et al., 1977:27).

The forty-nine offenders surveyed were serious habitual offenders. Together, they averaged twenty-one years of criminal activity from the time of their first reported offense up to their current incarceration. Almost 50 percent of that time was spent incarcerated. Although these offenders averaged only eighteen offenses per year of street time, the report rates underestimated individual offending frequencies of active offenders because they failed to exclude offenders who were not criminally active in an age period.

This pioneer study produced two of the principal findings to come out of criminal career research. First, its findings suggested that these offenders were largely unspecialized, engaging in a variety of crime types at any one time. "Most impressive was the extreme diversity in criminal activity shown by this small sample of recidivists, when the selection criteria had biased the sample toward homogeneity" (Petersilia et al., 1977:28). Secondly, it was found that the distribution of individual crime rates was skewed heavily, with most offenders committing crimes at a fairly low rate and only a few committing crimes at high rates (Petersilia et al., 1977:29). These two findings were subsequently confirmed by a

Table 2.1. Average Offense Rate per Month of Street Time

Offense Class	Juvenile Period	Young Adult Period	Adult Period	Entire Career
Violent[a]	0.10	0.16	0.20	0.15
Safety[b]	1.15	0.43	0.24	0.49
Nondrug[c]	2.37	0.92	0.38	0.99
All[d]	3.28	1.52	0.64	1.51

Source: Petersilia et al., 1977, p. 27.

Note: Figures obtained by dividing all offenses reported by the total number of months at risk for the entire sample.

[a]Violent offenses include robbery, aggravated assault, and rape.

[b]Safety offenses include burglary in addition to the violent offenses.

[c]Nondrug offenses include auto theft, purse snatching, theft over $50, and forgery, in addition to safety offenses.

[d]Total offenses include drug sales in addition to nondrug offenses listed above.

much larger survey designed to include all types of male prison inmates, involving 624 prisoners selected from five California prisons to represent the entire male prison population in custody level, age, offense, and race (Peterson and Braiker, 1981).

A second large-scale survey of inmates carried out by Rand researchers, which is described in Marquis (1981), Chaiken and Chaiken (1982), and Peterson et al. (1982), included 2,190 male inmates from prisons and jails in three states (California, Michigan, and Texas). This study, too, found that offense-rate frequency distributions were extremely skewed toward the high end of the range. For all types or combinations of offenses, the majority of subjects who reported committing the offenses reported doing so at fairly low rates. However, for every type of crime, a small fraction of offenders reported crimes at much higher rates.

This large variation in individual offense rates raised the question whether high-rate offenders could be distinguished for selective sentencing purposes. If this could be done, the amount of crime prevented through incapacitation might be increased by extending the time served in prison by the relatively small number of high-rate offenders, and decreasing the time served by the majority of low-rate offenders. In *Selective Incapacitation,* Peter Greenwood with Allan Abrahamse used the data from Rand's second inmate survey to support an argument "that selective incapacitation strategies may lead to significant reductions in crime without increasing the total number offenders incarcerated" (1982:xix).

From their analysis of subjects serving sentences for robbery or burglary, Greenwood and Abrahamse (1982:50–53) derived seven items that appeared to be associated with individual rates of burglary and robbery:

1. Prior conviction for the same type of offense
2. Incarceration for more than 50 percent of the preceding two years
3. Convictions prior to age sixteen
4. Having served time in a state juvenile facility
5. Use of hard drugs in the preceding two years
6. Use of hard drugs as a juvenile
7. Being employed less than 50 percent of the preceding two years

They used these seven items to construct a simple additive scale in which each item was given a value of 1 if true and 0 if not. Those who scored 0 or 1 on this scale were predicted to be low-rate offenders; those who scored 2 to 3 were predicted to be moderate-rate; and those who scored 4 or more were predicted to be high-rate (Greenwood, 1982:50–53).

Greenwood and Abrahamse argued that a sentencing policy that increased the length of time served by predicted high-rate offenders and reduced the time served by predicted low- and medium-rate offenders would reduce the robbery rate in California by about 15 percent and reduce the number of incarcerated robbers by about 5 percent. Under a policy of increasing the time served by all convicted robbers equally, it would be necessary to incarcerate 25 percent more robbers to produce the same 15 percent reduction in robberies (Greenwood, 1982: xix).

Greenwood and Abrahamse acknowledged, however, that their analysis was subject to several limitations and suggested that their work should be replicated in different sites, using prospective data obtained from both surveys and arrest histories. They added also that "the critical assumptions of the model should be tested" (Greenwood, 1982:xx). In the years that followed, considerable critical discussion surrounded the methodological and policy issues raised by their study.

The "incapacitation theory" that Greenwood and Abrahamse propounded was both similar to the approach of habitual-offender legislation and distinguishable from it: similar in searching for the high-rate offender as a candidate for long prison terms to achieve incapacitation; distinguishable because Greenwood and Abrahamse based the overall allocation of prison space on incapacitation criteria and thus reduced prison sentences for lower-risk offenders. The older habitual-offender approach left prison sentences in non-habitual-offender cases to be decided on other criteria. Thus incapacitative considerations had a pervasive impact on sentencing that had never been claimed for them earlier.

The report of the National Academy of Sciences Panel on Research on Deterrence and Incapacitative Effects, published in 1978, stated in relation to the incapacitative effective of imprisonment that

> a policy of selectively imprisoning the worst offenders (those who commit the more serious crimes and have a higher rate of committing crimes) has the potential of increasing the incapacitative effect.

The report began its recommendations for research into incapacitation with the words:

> Further, the critical development of incapacitation models is necessary to more accurately reflect variations in the patterns of individual criminal activity. To accomplish this, research should be directed at characterizing the patterns of individual criminal careers. (Blumstein et al., 1978:76, 78)

In 1983 a Panel on Research on Criminal Careers was convened

> to evaluate the feasibility of predicting the future course of criminal careers, to assess the effects of prediction instruments in reducing crime through incapacitation (usually by incarceration), and to review the contribution of research on criminal careers to the development of fundamental knowledge about crime and criminals.

The preface to the panel's two-volume report *Criminal Careers and "Career Criminals,"* published in 1986, notes that "many aspects of the work of the panel can be viewed as a follow-up to earlier work by the

Panel on Research on Deterrence and Incapacitative Effects (Blumstein et al., 1986:x–xi). In fact, the earlier panel had been established in part to evaluate crime reduction claims made on behalf of the Shinnar and Shinnar model, and the Panel on Research on Criminal Careers had a similar mission with respect to the Greenwood and Abrahamse's work on selective incapacitation.

Greenwood and Abrahamse's methods and results were scrutinized and criticized by others (e.g., Cohen, 1983, 1984; Spelman, 1984; von Hirsch and Gottfredson, 1984), who argued that reanalysis of the same data suggested that the original estimates overstated the effects of the proposed selective incapacitation. Because these researchers raised questions about the validity of Greenwood and Abrahamse's results, the panel commissioned a reanalysis of the original survey data and recomputation of the incapacitative effects estimated by Greenwood. The reanalysis resulted in various estimates of the effects of a selective incapacitation policy, all of the estimates substantially smaller than those produced by Greenwood and Abrahamse and all involving substantial increases in prison population (Blumstein et al., 1986:134).

Moreover, even those smaller estimates were, according to the panel report, "optimistic." In actual fact, the report went on, several factors would diminish the crime reduction achieved:

- The incapacitative effect will be less if the offenses of incarcerated inmates persist in the community, perhaps because the inmate is replaced by a new recruit or because incarceration of some members of offending groups does not disrupt the groups' crimes.
- The predictive power of the scale used to distinguish offenders may diminish as the scale developed on inmates is applied to a broader and potentially different population of all convicted offenders, especially in a different jurisdiction with different existing criminal justice decision practices.
- Self-reports of predictor variables will probably not be usable and will probably be distorted if they are used, and official records of those variables may have less predictive power because of record inaccuracies or gaps.
- The offending rates of low-rate offenders may increase in response to the shorter sentences imposed on those individuals under a selective policy because of a decreased deterrence effect.
- Finally, experience with efforts to reduce criminal justice system discretion suggested that the system may adapt to prediction-based decision rules in ways that reduce the considerable sentence disparity of eight years of predicted high-rate robbers and one-year terms for all others.

The magnitude of these effects, the report admitted, was "unknown at present," but they "would all operate to reduce incapacitative effectiveness" (Blumstein et al., 1986:135).

In *Selective Incapacitation,* Greenwood and Abrahamse reported selective incapacitation effects for a range of increases in prison terms for high-rate offenders while limiting the jail terms of other incarcerated offenders to one year. In particular, they predicted that a policy that increased the length of time served by predicted high-rate robbers in California while reducing the time served by predicted low- and medium-rate robbers would reduce the robbery rate by about 15 percent and the number of robbers incarcerated by about 5 percent (Greenwood, 1982:xix, 79).

The panel considered a variety of estimates of the effects in California of a highly selective incapacitation policy that doubled prison terms to eight years for inmates classified as high-rate robbers while imposing one-year jail terms on all other robbery inmates. It reported that

> [d]epending on assumptions about career length, the policy is estimated to reduce the number of robberies by adults by six to fourteen percent, with effects on robbery inmates ranging from essentially no change to a thirty-eight percent increase. These alternative crime-reduction estimates are all smaller than the fifteen to twenty percent decreases in adult robberies estimated by Greenwood. (Blumstein et al., 1986:134)

Subsequently, Greenwood acknowledged that subsequent reanalyses of the Greenwood and Abrahamse self-reported offense data suggested that their initial estimates were "overly optimistic" (Greenwood and Turner, 1987:49). In particular, William Spelman argued that a number of adjustments to those data and to the Greenwood-Abrahamse incapacitation model were necessary, including truncating extreme values and taking into account expected residual career length, the participation of multiple offenders in many offenses, and the engagement by most offenders in many different crime types. According to Spelman, these adjustments produced estimated incapacitation effects that were only about 5 percent as large as those estimated by Greenwood and Abrahamse (Spelman, 1986:vi–vii).

Finally, Greenwood and Turner undertook a study designed to determine "whether the seven-item scale could be used to predict the individual offense rates of convicted offenders and to investigate the loss of information that could be expected to result from using recorded arrest rates rather than self-reported offense rates" (1987:vi–vii). The study investigated how accurately the scale could predict individual postrelease arrest rates of two groups of California offenders: approximately 2,700 young men released from two California Youth Authority (CYA) facilities during the early 1970s, and approximately 200 former prison inmates who had responded to the 1978 Rand Inmate Survey (RIS) and had been released for at least two years.

It was found that for both samples the scale was only about half as accurate in predicting follow-up arrest rates as it was in predicting retrospective self-reported offense rates. The highest relative improvement over chance achieved with the scale was only 24 percent. Although the average arrest rates experienced by the predicted high- and low-rate groups in the CYA sample differed by more than a factor of two, the difference in average arrest rates between the predicted high- and low-rate RIS groups was only about 30 percent. Moreover, it was also found that there was little or no apparent correlation between individual offense rates and individual arrest rates. Greenwood and Turner concluded that

> [t]he poor predictive accuracy and modest differences in average arrest rates of the groups categorized by the scale do not appear to justify the large differences in sentence length for offenders in different categories that would be necessary to achieve significant selective incapacitation effects, at least for the types of chronic offenders studied here. (1987:48)

The last group of publications we note here deal not with a particular type of incapacitation but with a different dimension for the measurement of the effect of prison incapacitation, in terms of pecuniary costs and benefits. These publications include the writings of National Institute of Justice economist Edwin W. Zedlewski (1985, 1987), Princeton professor John J. Dilulio, Jr. (1989, 1990, Dilulio and Piehl, 1991), and Assistant Attorney General Richard B. Abell (Abell, 1989), each of whom has made contributions to what has been called the "debate [that] is raging over the cost-effectiveness of imprisonment" (Dilulio and Piehl, 1991:28).

The new dimension added by this set of papers is the attempt to translate incapacitation estimates from the Rand studies into a set of dollar costs of crime that are then compared to the dollar costs of imprisonment. This literature, discussed in detail in chapter 7, begins with the claim that each additional year of prison confinement reduces crime rates by a unit volume of 187 at a total social cost saving of $430,000. Unlike the optimistic claims for the Shinnar and Shinnar model and for selective incapacitation, no research panel was appointed to scrutinize these postulated benefits. As demonstrated in chapter 7, no panel of experts is required to expose the deficiencies of this approach.

II. Incapacitation as a Literature

Many of the analyses mentioned in this chapter will be critically reviewed in later chapters, when the specific issues they address are examined in detail. In this section, we wish to discuss briefly the material we have surveyed as a whole—as a literature dealing with the topic of the restraining influence of incarceration as it has evolved over time. The features of the incapacitation literature we wish to emphasize relate, first, to the absence, until quite recently, of sustained analysis and, second, to the episodic, dialectic, noncumulative, and nonempirical character of the materials that have accumulated over the past two decades.

However, to concentrate attention on what has been published over the past two decades would be to miss the most stunning point that can be made about the incapacitation literature—that is, the absence of any detailed analysis of incapacitation issues for more than 150 years, when the prison was becoming a dominant institution in the Anglo-American world. In 1802 Jeremy Bentham discussed in some detail the nature of imprisonment as incapacitative without, however, ascribing to this aspect of imprisonment any more than a subordinate role in the significant functions of imprisonment (1843:174). The matter was then totally ignored until the early decades of the twentieth century, and then considered only in connection with habitual offenders for whom deterrent and reformative initiatives had repeatedly failed. Interest in incapacitation was restricted to its application to habitual or especially dangerous offenders for another three-quarters of a century.

Neither intellectual nor technical difficulties can explain this legacy of inattention. Both the function and limits of incapacitation were as obvious in Bentham's time and as easily conceptualized by him as would be the case 175 years later. Indeed, the conceptual basis of incapacitation theory, characterized by Bentham as "as firmly laid in school logic as could be wished" (1843:83), was not significantly complicated in the 1970s and 1980s, when modern scholars addressed it.

Nor was any qualitative or methodological breakthrough associated with, or required for, empirical research into incapacitative effects. Estimates of the number of offenses committed by different types of offenders and arguments about the length of the criminal careers of individual offenders, could have been formulated much earlier, as finally happened in the 1970s and 1980s.

The scholarly literature that finally emerged in the 1970s was both episodic and dialectic. It is episodic because it consists of clusters of analyses concerned with discrete methods and claims during relatively brief periods. For example, the Shinnar and Shinnar paper was published in 1975, much discussed and criticized for a period of four or five years, then relegated to footnote status during the 1980s and early 1990s and to occasional listing in references. The research relating to claims about selective incapacitation similarly enjoyed wide attention and a short half-life during the period from the publication of Greenwood and Abrahamse's *Selective Incapacitation* (1982) until the publication of *Criminal Careers and "Career Criminals"* by Blumstein et al. in 1986 and Greenwood and Turner's *Selective Incapacitation Revisited* in 1987.

We may be in the middle of a third such episodic set of publications begun in the late 1980s that are organized around so-called cost-benefit studies of imprisonment (e.g., Zedlewski, 1987; Abell, 1989; Dilulio, 1990; Dilulio and Piehl, 1991), with the pendulum swinging back from selective incapacitation to collective or aggregate incapacitation as a function of imprisonment. In this case, claims about the cost effectiveness of prisons have provided the novelty of approach that seems to be required in order to spark episodes of renewed scholarly interest.

The dialectic quality of many papers dealing with incapacitation is one reason for the episodic pattern of publication. Most of the individual publications can be sorted into those making claims for a particular approach and those criticizing or suggesting modifications to the original claims. Once a specific claim has been deflated, the authors' approach is relegated to dead-end status in the literature. Published discussion seems to take the form of thesis, antithesis, and then silence. Thereafter, renewed interest, renewed publication, and renewed research activity seem to await the appearance of new claims or another novel thesis rather than further development of the ideas and perspective that provided the impetus for the previous "episode."

Thus, for example, the cluster of publications arising from the Shinnars' 1975 paper can be neatly divided into those supporting their claims and those criticizing and trying to refute them. This kind of dialectical structure centers the debate on particular claims about the effects of incapacitation rather than on underlying substantive issues. The debate is so centered on specific claims that for the scholarly interest generated by the studies inevitably tends to evaporate when claims are refuted or deflated. Thus, the first of the National Academy of Science Panel's reports, which undermined the claims of the Shinnars' study, tended to close down the field of inquiry rather than to stimulate efforts to study the effects of incapacitation in different ways. The same cycle of claim–critique–silence occurred in the wake of the publication of *Selective Incapacitation* (Greenwood, 1982).

As long as the academic literature on incapacitation is organized around specific claims, there is a pronounced tendency for that literature to become noncumulative. Each new claim is phrased in a distinctive vocabulary, and the critical reaction seems to have the effect of throwing the whole question back to naught rather than adding knowledge by a process of accretion. In this way, the incapacitation literature is noncumulative because each new dialectic episode seems to start from scratch rather than learning from, and building on, earlier discourse. At least three times in twelve years we have witnessed the publication of findings that exhibit that "start from scratch" quality: Shinnar and Shinnar in 1975, Greenwood and Abrahamse in 1982, and Zedlewski in 1987.

Another problem associated with the dialectical quality of the incapacitation literature is that undue amounts of effort and resources have been devoted to essentially critical activities. Two reports by the National Academy of Sciences Panel (Blumstein et al., 1978; Blumstein et al., 1986) represent an extensive investment of intellectual resources concentrated on the task of returning a field of inquiry to ground zero. Yet those activities involved a larger number of scholars by far than have been supported in federally sponsored *empirical research* on the subject of incapacitation. The lack of sponsored research on the issue must have been a disappointment to these scholars, whose reports called for support of much more research than subsequently occurred (Blumstein et al., 1978:81; Blumstein et al., 1986:198–99). Yet their concentration on the particular claims of published studies probably contributed to the perception that major research was not necessary once particular claims had been refuted.

One final characteristic of the published material on incapacitation is that much of it is unconnected to original empirical research. Only the Rand studies that led to Greenwood and Abrahamse's *Selective Incapacitation* generated empirical data of substantial dimensions. The inmate surveys were discontinued by 1980 and then replicated in a series of further studies in the 1990s. All the other incapacitation claims are based on secondary statistics or on extrapolation from data sets produced by others. Nor, with their critical orientation, did the panels of the National Academy of Sciences venture into substantial data gathering. We are not suggesting that the collection of fresh data necessarily advances knowledge, but analyses and debates that proceed for years without empirical testing or the development of new data sets are not the stuff of which cumulative progress in the social sciences is usually made.

3

Elements of Theory

To devote a chapter to theoretical approaches to a subject like incapacitation would seem to many to be a needless conceit. For the mechanism
of incapacitation seems like simplicity itself, needing only what Jeremy
Bentham called "schoolboy logic" (1802:183) for its explication, and
requiring models of implementation that would appear to be too simple even to merit that description. Where are the complications or
potential surprises that call for the skills of a theoretician? Why might
the model builder be worth his hire?

Certainly the behavioral mechanism at the heart of incapacitation is
much simpler than the processes that have been postulated to reduce
crime by means of deterrence or rehabilitation. Prevention through
incapacitation need involve no technique designed to persuade offenders or to change their characters, but can operate simply by restricting
the capacity of individuals to express their own preferences. In this
important sense, the machinery of incapacitation does not have any
moving parts.

Estimating the crime prevention benefits to be obtained by restraint
is in fact a complicated process and one that requires detailed knowledge of the psychology of both the potential offender and the potential
victims of crime. Though plausible theories of incapacitation need not
be hypercomplex, both careful modeling and rigorous testing are necessary before useful and reliable information about the net benefits of
restraint can be obtained.

This chapter researches the literature of recent years for some of the
elements that must be included in a plausible model of imprisonment as
an engine of incapacitative prevention. The first section analyses the
impact of various forms of restraint on the capacity of individuals under
restraint to commit offenses. The second section discusses the degree to

which high-risk offenders are selected for prison and the impact of this on the crime prevention realized by policies that produce additional prisoners. Finally, the third section considers when and how changes in the behavior of persons subject to restraint might produce lower crime risk in the communities from which they have been withdrawn.

I. Individual Effects

The *individual impact* of incarceration refers to the number of criminal acts that an individual would have committed had he or she not been restrained from doing so by reason of imprisonment. If Smith is confined in prison for a year and would otherwise in that time have performed 100 drug sales in the community from which he has been removed, the individual effect of his being restrained is the prevention of 100 drug sales.

However, 100 fewer drug sales will not necessarily take place in the community from which Smith has been removed. The *community impact* of Smith's removal is the net effect of his absence on drug sales in the community setting. This could be considerably less than 100 sales if Smith's associates continue to handle his sales or other dealers step in to supply the demand he has been fulfilling. The community effect of Smith's incarceration, however, could also be considerably larger if the 100 sales he would have performed supplied a pipeline of retail distributors, and many of the persons he supplied were unable to find alternative sources of supply. Thus, it is rarely safe to assume that individual- and community-level incapacitation effects would be identical.

This example illustrates the extent to which the simplicity of incapacitation is illusory at both the individual and community levels. Though it is true that the mechanism of physical restraint does not require knowledge of an offender's particular psychology, one cannot estimate the number of individual crimes prevented by means of incapacitation without knowing a good deal about the individual subject to restraint. How much and what kind of crime would that individual have committed if he or she had not been restrained? How long would the pattern of crime have persisted? Obtaining reliable answers to that sort of question requires a good deal of knowledge of individual psychology.

Taking the measure of community incapacitation effects further requires knowledge of the social organization of crime in the community from which offenders have been withdrawn. To continue the drug sale example, it is necessary to ask whether the individual sales were a solo operation or part of a group offense. If the sales were part of a group effort, will the rest of the group desist from or persist in selling drugs if the individual is restrained? If the offender's collaborators desist, will others come forward to supply all or part of the illicit demand? If so, how much of that demand will be met? Although the incapacitation mechanism itself may be simple, measuring its effects in the community is far from a simple matter.

Obviously the social benefits of incapacitation are not to be found or measured within the walls of prison or jails but rather in the community from which the potential offenders have been removed. Discovering how many offenses imprisoned persons would have committed had they been at large is only beginning of the measurement of the net incapacitative effect of incarceration. The ideal experiment would involve the random assignment of identical communities to contrasting experimental conditions of high and low imprisonment. The difference in the volume of crime experienced in the respective communities would provide an appropriate measure of the difference in the degree of incapacitation achieved by the differential imprisonment policies. Furthermore, the difference in the volume of crime divided by the number of person years of incapacitation would provide a measure of the average contribution of each year of confinement to crime reduction. Though this is not an experiment likely to be carried out in a democratic country, it offers a standard of comparison for existing, less rigorous methods employed to estimate levels of incapacitation in community settings.

The Arithmetic of Individual Incapacitation

The simplicity of incapacitation theory is not wholly specious. The Panel on Research on Criminal Careers, which was convened by the National Institute of Justice in 1983, defined incapacitation as "the removal of a convicted offender from the community, usually through imprisonment, to prevent the offender from committing further crimes" (Blumstein et al., 1986:15). As Jeremy Bentham put it, "for a body to act in a place it must be there." Once removed "for a given time: he will neither pick a pocket, nor break into a house, nor present a pistol to a passenger . . . within that time" (1843:183). It is incontrovertible that an offender cannot commit crimes in the general community while he or she is incarcerated.

Moreover, when it comes to the question of measuring the incapacitative effect of imprisonment—that is, estimating the number of crimes avoided by removing an offender from the community—some of the components of such estimates of incapacitation effects are legitimately quite simple and not seriously disputed. Thus, the effectiveness of incapacitation as a crime control strategy clearly depends on the dimensions of individual criminal behavior, for example, how frequently offenders commit crimes and the duration of active criminal careers.

The basic model of the individual criminal career adopted by the Panel on Research on Criminal Careers is shown in figure 3.1, which displays the essential concepts of the individual criminal career in terms of the three primary elements of information: the frequency or mean individual crime rate, λ; the age at career initiation, a_0; and the dura-

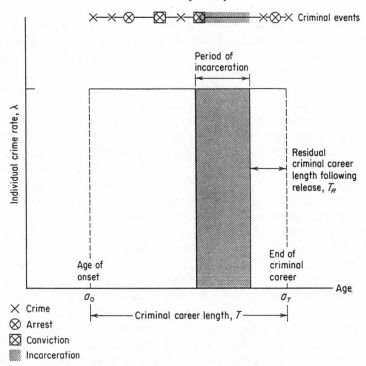

×—×—⊗——⊠—×—▨▨▨▨▨——×⊗× Criminal events

Period of
incarceration

Residual
criminal career
length following
release, T_R

Individual crime rate, λ

Period of
incarceration

Age of
onset

End of
criminal
career

Age

a_0 a_T

× Crime
⊗ Arrest
⊠ Conviction
▨ Incarceration

|←——— Criminal career length, T ——→|

Figure 3.1 An Individual Criminal Career. Data from Blumstein et al., 1986, p. 21.

tion of the criminal career, T. This is described in the panel's report as a
"minimum representation, which clearly omits many of the complexi-
ties of a real career." The offender is assumed to begin criminal activity
at some "age of onset" (a_0). Once begun, the offender continues to
commit crimes at a constant rate (λ) during any time he is not incarcerated. The offender's career ends when the last crime is committed, represented at age (a_T) (Blumstein et al., 1986:21–22).

What we refer to as individual incapacitation levels can be computed by multiplying the period of incapacitation by the rate of crime
that would have been committed by an individual but for his incapacitation. Computing the aggregate savings in crime prevented is then a
matter of completing all the individual calculations and adding the
totals into an aggregate value. These individual totals do not necessarily
correspond, as we have already noted, to the amount of crime prevented in the community over the period of incarceration, but obtaining these totals is almost universally regarded as a useful step in calculating incapacitation effects.

With some minor variations, this basic arithmetic formula provides
the underlying structure of the 1986 report of the Panel on Research
on Criminal Careers, Greenwood and Abrahamse's report on selective
incapacitation, and the model described in the Shinnars' article on

crime control by incapacitation (Blumstein et al., 1986; Greenwood, 1982; Shinnar and Shinnar, 1975). The principal difficulty involved in the computation is estimating the level of individual crime interrupted by incapacitative imprisonment.

Two limitations of the individual model of incapacitation that has been used to date concern the inadequacy of the simple model as a theoretical explanation of criminal behavior and as a testable and measurable hypothesis. The model of a criminal career presented in figure 3.1—a process that once initiated continues consistently at a fixed rate for a finite period and then suddenly ceases—is a paradigmatic computational convenience, but it accords with no known theory of criminal behavior or human motivation and resembles nothing so much as a battery-operated electric toy that is switched on to function at a fixed rate until its power source expires.

In fact, the assumption of a fixed lambda offender directly contradicts theoretical explanations of criminal behavior that are based on conceptions of rational choice, social contingency, or any other significant hypothesis about an offender's environment as an explanation of his behavior. A fixed rate of criminal behavior either contradicts most psychological or sociological theory about the causation of criminal behavior or assumes that psychological and environmental forces are held constant for very long periods of time. The inconsistency between the assumption of a "fixed lambda" protagonist and any known theory of crime causation has not been explained anywhere in the literature.

Though no necessary mischief is intended in the use of constructs like the fixed lambda that are artificial and also in conflict with other theory, there is a danger that the tidiness and symmetry of the picture—and the ease of computation based on the model—may lead users to assume that it reflects behavioral reality.

Another limitation of the individual model of criminal careers absent incapacitation is that many of the key dimensions of the model are not at the present time subject to reliable measurement. Detailed observation of large numbers of persons who commit crime would be necessary over long periods of time before it would be possible to assemble the sort of historical data that would cover the length, composition, and contingency patterns of offenders.

Rand researchers noted the artificiality of fixed-rate assumptions for criminal careers in the early 1980s, and a number of alternative methods of modeling crime commission rates were tested in the second Rand inmate survey (Rolph, Chaiken, and Houchens, 1981). More recent work by David Greenberg, as well as Daniel Nagin and Kenneth Land, has used arrest and conviction data from cohort studies to test the distribution of criminal career events over time and by age (see Greenberg, 1991, and Nagin and Land, 1993). However, this type of empirically disciplined analysis has not yet received the attention it deserves. Instead, model building and policy discussion are all too frequently carried on without any recognition of the need for information.

A model of individual offending that is much easier to justify as a way to organize data than as an attempt to explain criminal behavior turns out to have very little data to organize. Even those who recognize the great importance of unresolved empirical questions to incapacitation theory as an element of criminal control policy have been unsuccessful in launching sustained research (Blumstein, 1988). If ever a model was not meant to stand alone, it is the diagram shown in figure 3.1. Yet it clearly does stand alone under present conditions.

II. Selectivity, Predictability, and Diminishing Returns in Individual Incapacitation

One firm conclusion that may be drawn from all attempts to measure individual crime rate variations among convicted offenders is that the variation itself is quite substantial. Some offenders report very high rates of a variety of offenses; many others would be responsible for little or no crime if not incarcerated. For this reason both practitioners and theorists have attempted to fashion schemes to restrict the use of confinement but to maximize its productivity by selecting particularly high-rate offenders for penal confinement.

Thus in *Selective Incapacitation* Peter Greenwood and Allan Abrahamse claimed:

> In this report we have shown how incapacitation theory might provide a rational means for allocating scarce prison space. We have used self-reported data from prison and jail inmates to demonstrate that there is wide variation in individual offense rates and that the factors associated with higher rates of recidivism are also associated with higher rates of offending. Finally, we have shown that selective incapacitation strategies may lead to significant reductions in crime without increasing the total number of offenders incarcerated. (Greenwood, 1982:xix)

In the same vein, on the subject of incapacitation strategies, Alfred Blumstein and his colleagues stated:

> The recent large increases in inmate populations continue to place pressure on available prison and jail resources and to highlight the desirability of targeting incarceration more narrowly to achieve greater crime reduction from the available limited capacity. Increasing interest in a strategy of selective incapacitation has been further sparked by observations that individual offending frequencies differ widely among offenders, with a small number of offenders committing crimes at very high rates. . . . Such a skewed distribution suggests the possibility of increasing the crime reduction benefits from incapacitation by

selectively targeting incarceration on the small number of high-rate offenders. (1986:(I)128)

The impact of different selective incapacitation approaches will be examined in chapter 5. For now it is important to state with some care the relationship between the selection criteria, the predictability of particular criteria amongst a population of convicted offenders, and the efficacy of expanding the use of confinement as a means of further reducing crime.

Most proposals to select offenders either for imprisonment or for longer terms of imprisonment are concerned with both the volume and the seriousness of anticipated offenses. The correctional planner wishes to restrain convicted offenders who commit the most serious crimes as well as those who will commit the largest number of offenses if at liberty. These two criteria produce a simple matrix as shown in table 3.1.

The easily prioritized cases for a selective incapacitation schema are shown in groups A and D, exhibiting extreme values for both volume and crime seriousness. The group expected to commit a high number of the most serious offenses prove the most attractive candidates for any scheme of selective incapacitation. By contrast, group D represents the least attractive candidates: low in both volume and seriousness of offense, they produce the lowest incapacitation yield on both significant dimensions. Groups B and C represent middle-range cases: it is important to note that there is no a priori basis for choosing between groups for which we anticipate high numbers of relatively low-seriousness offenses if the offenders are left at liberty (group B) and groups for which a smaller number of offenses are expected but these are of a higher than average level of seriousness (group C). The choice between these two priorities is an important element in any program of selective incapacitation.

Any plan intended to choose selectively those who represent a higher than average risk of serious future crime from a large volume of convicted offenders must utilize known attributes of offenders to make discriminatory predictions regarding future offenses. The particular attributes currently used in criminal sentencing include the number of prior convictions and arrests of an individual; the seriousness of the offense that is the subject of the current conviction; the length of the period between prior conviction and punishment and the current arrest; and the extent to which the offender has strong lawful ties to the community, such as gainful employment, marriage, and family relationships.

Table 3.1. Criteria for Selective Incapacitation

		Seriousness of Anticipated Offense	
		High	Low
Number of		High	Low
Anticipated	High	A	B
Offenses	Low	C	D

Many of these attributes, such as the seriousness of current offense and prior criminal record, could justify differential penal treatment quite apart from their influence on an individual's propensity to commit crime in the future. But all of these factors can also be converted into predictors about future criminality if the offender is not restrained. The seriousness of the current offense might predict more serious future offenses either because the offender has been specializing in a serious crime and is thus more likely to repeat it or because a more serious infraction indicate a greater willingness to consider law violations of all kinds. Similarly, a longer record of convictions can be related to a variety of theories of greater future risk, the presuppositions ranging from habitual patterns of criminal behavior and the demonstrated failure of past penal measures, to an even more emphatically demonstrated commitment to antisocial norms.

Selection and Prediction

It is necessary to distinguish between processes that use predictions about the seriousness and amount of future criminal behavior and the wider range of processes that select higher-rate and more serious offenders for imprisonment, whether or not an explicit prediction informs the selection process. Whenever the police, prosecutors, judges, or parole authorities impose a larger measure of control because it is felt that the subject of the control is a likely candidate for recidivism, we observe the phenomenon of predictive sanctioning. The police officer decides to make the arrest because he or she thinks that the subject will otherwise continue to commit petty theft; the prosecutor refuses to reduce the felony charge to a misdemeanor; the judge rejects a nonimprisonment sanction; the parole board denies early release—all on account of worry about the possibility of future crime that leniency will risk.

If the predictions of dangerousness that criminal justice agents use to differentiate between groups of offenders are well founded, then the use of prediction will result in an imprisoned population with higher than average risk of future criminal behavior while the residual population of offenders not imprisoned will contain persons with lesser propensities to commit further crime. Sentencing policies that hope to achieve these risk determinations are *predictive*, whether or not the criteria employed actually result in assigning different risk groups to different punishments. By contrast, any sanctioning system that has the effect of confining convicted offenders with higher risks of future criminal behavior than others it does not confine is a *selective* system, whether or not the process is intended to base decisions on predictions regarding future crime.

Highly selective results can be produced by sanctioning systems when predictions of future criminal behavior play little or no formal role. If factors such as the length and seriousness of a prior criminal record, the swiftness of rearrest after a previous adjudication, or the gravity of the current offense are positively related both to the seriousness of the sanction to be imposed and to the likelihood and frequency of future crime, the resulting pattern of sanctions can be selective in effect without being preventive in motivation. Conversely, if prediction criteria do not operate as intended, a sanctioning system can be preventive in intention without being selective in effect.

We belabor this definitional distinction to ensure the clarity of the following hypothesis: the more selective in effect the system that determines which offenders are sent to prison, the greater the diminishing marginal return can be expected as a higher proportion of all convicted offenders is imprisoned. This can be illustrated as follows.

Assume a universe of twelve convicted offenders, six of them likely to commit ten offenses per year if unconfined and six of them likely to commit only two offenses per year if not. We will call the ten-offense group the *high-rate group* and the two-offense group the *low-rate group*. Assume further that current policy will produce the imprisonment of six of the twelve convicted offenders. Now if the processes that determine which offenders who go to prison are unselective, the current policy will produce an imprisoned population with roughly the same proportion of *high-rate* and *low-rate* offenders as in the total population of convicted offenders. In our example we can expect about three high-rate and three low-rate offenders to be captured by an unselective six-person sentencing policy. If three high-rate and three low-rate offenders are restrained for one year, a total of thirty six offenses, or six per imprisonment year, will be prevented. Moreover, adding to or subtracting from the proportion of offenders imprisoned will not cause any significant shift in the expected amount of prevented crimes per year of imprisonment: the number will remain at six.

If, however, the processes that determine imprisonment are highly selective, the amount of prevention achieved by each year of imprisonment imposed will vary with the proportion to be imprisoned. Assume that five of the six offenders imprisoned in our example case are members of the *high-rate group*, then the imprisonment of six will prevent fifty-two offenses, or 8.7 per imprisonment year. This is almost one-half again as efficient as the crime prevention associated with nonselective sentencing processes.

The price of this initial success is that each additional imprisonment in the selective system will prevent an average of only 3.3 crimes, less than 40 percent of the prevention associated with the first group of imprisoned offenders, and a little more than half the per capita crime prevented in the case of the nonselective sentencing of the same offenders. The more selective a sentencing system is in effect, the greater the diminishing marginal return as a higher proportion of eligible offenders are selected for imprisonment.

When existing criminal justice policies incarcerate a substantial fraction of a population of offenders, the expected returns from further incarceration may be inversely proportional to the efficiency of the current system in selecting high-risk cases for imprisonment. The more effectively the existing system operates, the less prevention per additional unit of imprisonment can be expected. By confining disproportionate numbers of high-risk cases, a selective sanctioning system is in effect working itself out of a job.

When diminishing returns are a likely phenomenon, it will of course be unwise, if the proportion of offenders to be incarcerated is increased, to generalize from the known characteristics and criminal propensities of the group currently incarcerated to the characteristics and behavior of the next group to be imprisoned. Under certain circumstances of selective imprisonment a particularly high rate of expected offenses among the already imprisoned should lead us to anticipate much lower levels of risk among the residual group of offenders not currently incarcerated.

There may be a particularly sharp drop where a very small number of offenders is responsible for an extremely high fraction of total crime. If, for example, one of the twelve offenders in our hypothetical case has an expectation of not 10, but 300, crimes if released, the greatest incapacitation effect that can be expected from penal policy will occur when that single individual is locked up. If current policies make it much more likely that such extremely high-risk offenders are already behind bars, the marginal return from further incapacitation will be much smaller than the efficiency per unit of imprisonment under the existing criminal justice system.

By contrast, when a system that chooses among offenders for imprisonment is totally nonselective, the characteristics indicative of propensity toward crime of those currently incarcerated are a good approximation of what the next group will look like. With nonselectivity both those currently imprisoned and the group that any policy change would capture should be close to representative of the general population of offenders. But even if the risk per act is uniform across a population of offenders, the more frequently criminal offenders are likely to be locked up first and some diminishing marginal returns can be expected. Only in nonselective systems is it safe to assume that the severity of sanction policies would have no effect on the characteristics of offenders who will be punished. The irony here is that only the inefficiency of the sanctioning system in selecting high-risk offenders can provide anything close to consistent incapacitation benefits as policies change. The sentencing system can either work consistently or well, but not both ways at once.

The statistical patterns that could produce significant diminishing marginal returns were recognized early in the Rand studies. Indeed the processes that lead to diminishing returns are the same as those that inspired the proposal for selective incapacitation (see Greenwood,

1982). Diminished returns from wider sampling in an offender population represent in one sense the flip side of selective incapacitation.
Whereas selective incapacitation schemes depend on improvement in
the degree to which imprisonment is rationed selectively to high-rate
offenders, diminishing returns are a byproduct of any selectivity that
criminal justice processes generate.

Though the processes that can generate diminishing returns have
long been observed, recognition of their importance for policy purposes is a more recent phenomenon. The degree to which a state's
criminal justice processes send a larger number of offenders to prison is
in fact identified as a major reason for lower median offense rates in a
New Orleans inmate survey by Miranne and Geerken (1991:507). The
offenders surveyed by the Colorado Department of Public Safety
reported lower rates of offending in 1989 than in 1986, a result that is
explicitly tied to increasing imprisonment and a changing population
in Colorado prisons (English and Mande, 1992). The policy impact of
these types of variation over time has been stated by Cohen and Canela-
Cacho:

> [S]tochastic selectivity, however, also results in decreasing marginal
> returns from incapacitation as increasing numbers of offenders are
> incarcerated. Just as high-rate offenders are over-represented among
> inmates, low-rate offenders are disproportionately found among the
> offenders who remain free. (1994:356)

Two points should be reiterated about selectivity in sentencing policies to underline the potential importance of diminishing returns. First,
the system need not be preventive in intention to be selective in effect.
There are good reasons to suppose that almost all sanctioning systems
that do not send the great majority of convicted offenders to prison
operate selectively. What we do not know is the degree of selectivity that
is operative in current penal systems.

Second, both the degree of selectivity in current systems and the
distribution of risk profiles among various groups of convicted offenders are variable. Since the individual effects of imprisonment are the
joint product of these two variables, they should vary both cross-section-
ally and over time.

If imprisonment policies play an important role in determining
what sort of offender is at the current margin between confinement and
release to the community, the marginal effect of incarceration will be
both variable and contingent. It will vary not only in relation to the
propensities of different offender populations in different communities
but also as a consequence of different policy environments. The
amount of crime to be prevented if ten more offenders are confined
may be as much a function of existing criminal justice policies as of the
criminogenic character of a particular community.

The contingent nature of incapacitation effects should negate the possibility of a narrow range of crime prevention estimates for changes in prison population in different times and places. Even if offender populations do not vary widely over time and between jurisdictions, confinement policies do vary substantially. In these circumstances there is little prospect of finding a single crime prevention estimate that can be generalized to different settings.

III. From Individual to Community Effects

If processes of restraint through incapacitation produce social benefit, this occurs in the community settings from which offenders have been removed. Since the community is where incapacitation aspirations reside, theoretical accounts of incapacitation are obviously deficient unless they provide a plausible account of how the restraint of individual offenders influences crime rates in the communities from which offenders have been withdrawn. Some of the many factors that require examination in relation to how incapacitative processes may have community impact are reviewed in this section under two broad headings: group processes and community environmental factors.

Incapacitation and the Criminal Group

A substantial amount of crime—up to one-half of street offenses such as robbery and burglary—is committed by groups of two or more offenders (Reiss, 1986:123–24). The social structures that produce group crime range from interdependent and role-differentiated criminal groups, all the way to informal street crime assemblies of short duration. One thing all these "co-offending" groups have in common is that they complicate the task of translating individual offense rate estimates into plausible estimates of the impact of imprisonment on the community.

The basic problem can be easily illustrated. Assume three offenders usually commit burglary together. If all three are apprehended and imprisoned, this will prevent a measurable level of crimes in most community environments. However, the crimes prevented will be one-third of the number that would be arrived at by adding together the individual crime rates estimated by the model shown in figure 3.1. Yet if that were the only problem, one could estimate group involvement and deflate individual crime avoidance estimates by some empirically derived group involvement factor (Reiss, 1986:156).

However, what if *one* of the three co-offending burglars is imprisoned? If the imprisoned offender was socially dominant or operationally necessary, his incapacitation might lead to desistance or reduction in criminal activity out of proportion to his share of the group's criminal activity. If, on the other hand, the two group members who remain at large persist in their previous pattern of criminal activity, only now as a dyad rather than a triad, or if they recruit a new third participant, the community-level incapacitation effect of the single imprisonment will be zero. For those situations where part, but not all, of a group previously involved in criminal behavior is imprisoned, we cannot know the aggregate crime prevention to be anticipated without some sense of the balance between desistant and persistent group responses.

The same self-report approach that has been used to estimate individual lambda rates might both profile the history and character of an individual's involvement in group crime and explore the balance between desistant and persistent responses when individual members of co-offending groups are removed. In theory, partial co-offender removal could produce a range of effects, from a dramatic reduction in group crime to net increases in community crime levels because of the recruitment of new offenders while those persons they replace are recirculated in the community after a relatively short period of time (Reiss, 1986:122). So the aggregate impact of group involvement on incapacitation is an empirical issue, and community-level incapacitation cannot be estimated without our being capable of answering that question. Yet this set of very important empirical questions has not to date become a priority feature of incapacitation research. The theoretical implications of group involvement in crime have been forcefully argued (Zimring, 1981; Reiss, 1986), but research designed to incorporate those implications into incapacitation findings has not been carried forward.

Where crimes have been committed by a group, the confinement of some but not all members of the group may be seen as an instance of *incomplete incapacitation,* but merely knowing the proportion of a group that has been confined cannot tell us the degree to which the group itself has been disabled in its potential for crime. Assuming that removing one of three collaborating burglars from a community will reduce the group's burglaries by one-third has arithmetic charm but no logical foundation.

Incidentally, one fruitful way of looking at the substitution problem in situations like drug sales would be to take victimless crime as a paradigm case of incomplete incapacitation. The drug sale used as an illustration earlier in this chapter involves a criminal collaboration between a seller and a buyer. Lock up the seller and the buyer may seek out another. But the buyer's demand for a criminal collaborator may not necessarily be any stronger nor any more predictable than that of a burglar suddenly deprived of a specialist accomplice or a recreational robber bereft of companionship in crime.

The Societal Environment of Crime

Just as the reaction of a co-offender group can determine whether individual-level incapacitation is translated into lower crime rates in the community, a number of characteristics of the social environment from which offenders are removed will influence the extent to which the incapacitation of individuals can result in lower community crime rates. Are there potential or actual offenders at large with the necessary skills to commit the offenses that would substitute for those the imprisoned offender would have committed? Are the criminal opportunities this offender would have exploited known or easily discovered by others? Are the risks and benefits associated with those crimes in this community setting attractive to other potential offenders?

In the case of co-offenders the mediating alternatives have been referred to as the *persistence* or *desistance* of a criminal group when one of the offenders has been removed. With respect to environmental influences the alternative patterns to be contrasted are *substitution* and *extinction*. Substitution occurs when others commit the offenses a prisoner would have engaged in had he or she been at liberty; extinction is the pattern whereby the removal of the offender from the community prevents in the behavior he or she would have engaged in from occurring. No attention has been previously devoted to the systematic explanation of factors tending toward substitution or extinction; the list we provide is both preliminary and incomplete.

The Market Metaphor

Among the factors that bias community-level responses toward substitution, the most prominently mentioned is the existence of a market for illicit goods and services. An explicit market for criminal services in which persons bid for the performance of criminal acts obviously facilitates substitution of the labor or services of incapacitated offenders. Two clear-cut examples of explicit markets are the sale of illegal goods and the commanding of illicit services. The sale of drugs, prohibited weapons, and other contraband are examples of the former. Examples of the latter include instances in which the putative victim is a willing purchaser, as in the case of prostitution, and also cases in which illegal services are purchased or recruited for purposes of predation against third parties, such as murder for hire and fencing schemes. The more explicit and the more widely known the market, the more likely it seems that substitution will nullify the preventative thrust of individual incapacitation. This prediction should hold equally for both victimless and predatory crime if market organization is similar.

It is important to emphasize that the existence of markets for criminal goods and services does not inevitably entail the total nullification of incapacitative effects. Such markets are rarely, if ever, perfect, and what economists refer to as "search costs" (see Moore, 1977) for both buyers and sellers can be sufficiently substantial to ensure a considerable measure of extinction of criminal behavior in the community as a result of the incapacitation of a particular criminal supplier. It seems likely that the closer the social and economic circumstances resemble auction markets the more likely it is extensive substitution will occur.

Another important consideration is that marketlike environments and circumstances promote substitution because they possess certain characteristics that influence its occurrence. Thus it is worthwhile to analyze separately what it is about societal environments that makes them marketlike in their capacity to facilitate criminal substitution in relation to certain types of crime.

Predisposing Conditions

Four conditions seem plausibly to be linked to substitution rather than extinction when previously active offenders have been incapacitated. We phrase the hypotheses in this list to predict the promoting of substitution as an expository convenience. The converse of the conditions listed should predict lower rates of substitution or extinction.

1. *Discrete, Fixed, and Limited Crime Opportunities.* If the opportunities to commit crime are unlimited and undifferentiated, the removal of one or more active offenders should have no influence on the number or kind of offenses committed by others because the opportunities and risks facing offenders do not change. By contrast, when the number of discrete criminal opportunities that are considered particularly attractive or accessible is limited, the prospect of substitution is more likely. In a world in which six prostitutes are soliciting only five customers on a street corner, the incapacitation of one prostitute will probably lead to complete substitution by the other five. On the other hand, by reducing demand, the removal of one potential customer in this case will produce a reduction in completed prostitution offenses of 20 percent.

Situations with discrete and fixed criminal opportunities are by no means restricted to victimless crimes. In some environments, the number of "drunk rollings" of public inebriates is likely to be a function of the number of intoxicated persons in vulnerable public circumstances who look as though they have something worth stealing, rather than a function of the number of potential offenders willing to roll drunks. If a restricted number of highly vulnerable commercial establishments represent particularly attractive targets for potential offenders, the incapacitation of an offender who specialized in exploiting those targets will be more likely to lead to substitution by other predators than where the number of targets is great and no pronounced vulnerability differentiates a small number of the potential targets.

2. *Large Numbers of Potential Offenders.* A second factor predisposing a community toward high levels of substitution is the existence of a large number of potential offenders. If a community has very few potential offenders of a particular kind—prostitutes, burglars, rob-

bers—the removal of active offenders is less likely to be counterbalanced by substitute activity by others than if the number of would-be offenders is large. If a community's only prostitute were locked up, this would be far more likely to reduce the number of acts of prostitution than in the hypothetical situation of the six prostitute–five customer example given above.

The factors that might restrict the supply of potential offenders include the lack particular skills required to perpetrate an offense; moral barriers that reduce the number of persons willing to attempt particular crimes; and high detection and punishment risks that incapacitate a large proportion of potential offenders and deter some of the remainder. Skill requirements limit criminal participation in many fraud schemes, counterfeiting, and the robbery of well defended targets. Moral barriers significantly reduce the number of potential participants in many types of crime and interact with particular skill levels or access requirements to reduce potential offenders to small numbers in, for example, national security espionage and in other criminal opportunities of restricted access. High punishment likelihoods are also associated with reducing the number of potential offenders available as substitute offenders in some settings. The converse condition is one in which large numbers of potential offenders are available to take advantage of a restricted number of particularly attractive opportunities.

3. *Widely Known Criminal Opportunities.* All other things being equal, the more widespread information is about criminal opportunities, the greater the chance that incapacitation will produce substitution. A case in point is the incapacitation of individuals or groups to whom particular criminal opportunities are a trade secret known only to them. Unless others discover the opportunities that those incarcerated were exploiting, their incapacitation should lead to the extinction of their pattern of criminal activity during the period of their incarceration. Blackmail is one example of a secret known only to the offenders and their victims. When particular criminal opportunities are widely known, in the jargon of economics, the "search costs" for potential offenders are lower. Widely advertised vulnerability to criminal predation, as in the case of the public inebriate mentioned earlier, will be more likely to attract substitute predators rather than to result in extinction.

4. *The Risk Environment of Crime.* A final influence on the extent of substitution is the risk of apprehension and punishment associated with a specific type of crime in a particular community. Other things being equal, the greater the chance that an individual committing the offense will be apprehended and incapacitated, the smaller the likelihood that the incapacitation of one offender will result in substitute criminal activity by someone else in the community. The reason for this is that high risks of apprehension and punishment can be expected to reduce the residual number of potential offenders at large, either by way of general

deterrence or by incapacitation. The arithmetic of the prediction in relation to incapacitation can be simply stated.

Assume two communities, A and B, each of which begins with a population of ten potential burglars and with the same rates of burglary, but with different apprehension and punishment risks. Eventually the high-risk community A has seven burglars in jail and three remaining on the street. The apprehension of another burglar would then reduce the number of burglars at large by 33 percent and would leave only two potential sources of substitution among active burglars in the community. In the low-risk community B, however, only five out of the ten burglars were apprehended and incarcerated in the same period of time so that the relative reduction in the street criminal force achieved by the incarceration of one of the remaining five on the street would be 20 percent. Four offenders would be left on the street as potential substitutes, and the chances of substitution would be correspondingly higher. The power of this prediction depends on there being a relatively small pool of potential offenders with a relatively high degree of vulnerability to apprehension and incapacitation.

Where the incapacitation process works cumulatively to reduce substitute criminal activity in a community, the cumulative effectiveness will compensate to some degree for the tendency discussed in the previous section for high-risk environments to imprison a larger proportion of offenders with relatively low individual criminal propensities. In our community comparison, the first three or four offenders caught and punished in community A are likely to have engaged more frequently in burglary and thus to have exposed themselves more often to the risk of apprehension than their less active peers. The fifth, sixth, and seventh burglars apprehended are likely to have lower rates of offending and therefore to offer diminishing prospects of individual incapacitation impact.

This diminution could be offset to some extent by the reduction in the number of offenders at large producing lower rates of substitution. However, though this prediction depends on a Shinnar-style arithmetic calculation, it is a point that has not been explored in the available literature on incapacitation. This may be because the problem of substitution has been largely ignored rather than being explicitly incorporated into current models of incapacitation.

Estimates and Measurement

Because of the manifold possibilities for group-process and community-level influences, even if individual effects could be estimated with precision, the level of crime reduction experienced in the community from imprisonment could not be determined using individual incapacitation rates. One obvious way to deal with this problem is to assume particular discounts to estimate group-process and substitution effects (Spelman,

1994:58–61. However, the impact of these effects is too indeterminate to support any hypothesized mechanism for converting individual-effect estimates into community-effect estimates. Perhaps when more is known about the factors that influence persistence and substitution, a conversion exercise may appear less fanciful.

At the present time, however, it is necessary to study the response to incapacitation in the community both to generate data relating to environmental and group-process effects and to search for plausible estimates of the net impact of shifts in incapacitation on community crime rates. Comparing incapacitation estimates derived from community data with data derived from prisoners' self-reports of individual crime rates may be one way of obtaining some sense of the potential extent of persistence and substitution.

We do not wish to minimize the difficulties confronting researchers who study fluctuations in community crime rates for evidence of imprisonment policy influences. As discussed in chapters 5 and 6, such study is a formidably difficult undertaking. It is the lesson of this chapter that such community-level measurements are an indispensable part of the assessment of the effect of incapacitation policies. That no attempts at community-level assessment can be found in the literature on incapacitation highlights the pressing necessity for the design and conduct of such research.

4

The Jurisprudence of Incapacitation

This chapter deals with the jurisprudence of imprisonment for the purpose of incapacitation, that is, with the justification and limits in the criminal law of prison sentencing based either on special findings of dangerousness for particular offenders, or on a general theory that prison sentences should be administered with the primary purpose of restraining future criminal behavior. The four-part organization of the chapter is a response to the somewhat curious pattern of coverage that characterizes the literature dealing with these topics.

Section I notes the contrast between the jurisprudence of special incapacitation, a topic that has been extensively discussed in academic criminal law, with jurisprudential issues concerning collective or general incapacitation strategies, a topic about which very little has been written. Section II surveys the principal issues surrounding the justification and limits of special incapacitation that have been addressed in modern writing. Section III discusses whether the problems presented by a collective incapacitation strategy are significantly different from those generated by a strategy of special incapacitation; whether any differences between general and special incapacitation require different patterns of justification for their use in the criminal law; and whether these differences also call for different limitations on the two types of incapacitation as a purpose of prison sentences. Section IV then outlines the principles we feel should properly guide and limit a system of prison sentences that depends in large part on restraint as a general justification for its implementation.

I. A Study in Contrasts

An extensive literature in American criminal law debates the extent to which the imposition of prison sentences should be decided on the basis

60

of judgments about the likelihood that individual offenders will commit further crimes if not confined or continued in prison. For a period of just under a decade, from the early 1970s onwards, this was one of the most prominently addressed issues of criminal sentencing, evoking major contributions from Norval Morris, Andrew von Hirsch, Marvin Frankel, Michael Tonry, and many others (Morris, 1974, 1976, 1984; Morris and Miller, 1985; von Hirsch, 1976, 1985; Frankel, 1973). By contrast, there has been little discussion of the justice of basing a whole system of prison sentences, not on the need for restraint in individual cases but on the desirability or necessity for imprisonment of most offenders, to prevent further offenses. Possibly the difference in coverage is based on the judgment that special and general incapacitation are significantly different penal practices, but we know of no sustained relevant discussion.

In fact, the unbalanced emphasis in academic discussion on special incapacitation seems to have been related in both time and context to declining faith in the rehabilitative ideal. Predictions regarding individual dangerousness were an implicit part of the justification of judicial and parole board discretion in choosing between prison and other sentences and in setting the duration of prison terms. These exercises of discretion seemed ripe for reexamination in the 1970s. General incapacitation, by contrast, was a strategy associated not with the rehabilitative justification for imprisonment but rather with the systems of criminal justice that rose to dominance by a process of elimination after the decline of the rehabilitative ideal. The underlying assumptions of the new system have not yet been subjected to the same critical analysis and profound skepticism that the rehabilitative rationale encountered in the twilight of its reign as the dominant justification for imprisonment.

We intend to refer to the relatively substantial body of work analyzing and discussing the topic of special incapacitation as a basis for predicting how analysts might decide the crucial questions surrounding the policies of general incapacitation.

II. The Jurisprudence of Special Incapacitation

The late 1970s debate about special incapacitation and prison sentences was principally concerned with three questions:

1. whether extra imprisonment for those believed dangerous presented special problems of unfairness that did not arise when other principles like desert and deterrence governed the selection and length of prison sentences for convicted offenders;
2. whether particular problems posed by predictions of dangerousness required that consideration of whether an individual offender was at risk of committing further crime in the community should not be a factor in the determination of punishment;

3. whether there were safeguards short of prohibition on the use of predictions of dangerousness in sentencing that could be equitable and effective.

Specific Problems of Special Incapacitation

A variety of different objections have been raised in relation to predictive sentencing, but not all have been the subject of equal concern. What has been seen as the central vice of predictive sentencing is the supposed unfairness of punishing persons not simply for what they have done but also basing such decisions on judgments about what they might be likely to do in the future. Some critics characterize the practice as punishing for future crime and regard that as unjustified because the criminal behavior has not yet occurred. Indeed, one critic, Andrew von Hirsch, would regard the escalation of punishment because of the risk of future crime as not morally justified, even assuming that accurate prediction of the occurrence of that crime were possible. "Prospective considerations—the effect of the penalty on the future behavior of the defendant . . . should not determine the comparative severity of penalties" (von Hirsch, 1985:31). Others have pointed to the uncertainty that surrounds predictions of future criminality and emphasized the injustice of confining individuals in the erroneous belief that they will commit criminal acts in the future.

This "false positive" problem was the focus of concern for Norval Morris in *The Future of Imprisonment:*

> Even when a high risk group of convicted criminals is selected, and those carefully predicted as dangerous are detained, for every three so incarcerated there is only one who would in fact commit serious assaultive crime if all three were released. . . .
> . . . [A]s a matter of justice we should never take power over the convicted criminal on the basis of unreliable predictions of his dangerousness. (1974:72–73)

Presumably Morris would not regard some extra period of imprisonment as unjust if the predictions were accurate, whereas von Hirsch would regard as objectionable any increase in prison sentences on account of criminal behavior not yet committed, no matter how accurate the prediction.

These two different views of the special problem of predictive punishment were reflected in different judgments about which motives for imprisonment were morally justifiable. Not favoring general deterrence or incapacitation per se, von Hirsch felt that the only justification for the imposition of a sentence of imprisonment on an offender was the moral culpability of the offender's own behavior. Morris, by contrast, approved the escalation of punishment for the purpose of general deterrence as long as a sentence so determined did not exceed a range of deserved punishment. However, he did not, at least in his earlier writings, allow the extension of prison sentences within that range because individuals were believed dangerous.

The distinction between desert and deterrence drawn by Morris was concerned with the question of "false positives," and although the distinction may highlight a practical political problem, it is in fact logically unsound. The logical problem is that imprisoning 100 burglars because two-thirds of them will commit future crime if they are not restrained only involves false-positive predictions for 33 out of the 100 burglars. But at the same time, Morris approved "a general deterrent justification for the imposition of a prison sentence," mentioning in this connection "the appropriateness" of "imprisoning federal tax offenders" (1974:76, 79). Yet the imprisonment of 100 physicians for income tax evasion to achieve general deterrence, that is, to secure compliance with the tax laws by hundreds of other physicians as a result of that punitive example, appears to base the justification of punishment on moral grounds just as attenuated as those underlying the imprisonment of the "false positive" burglars. In both cases, the imprisonment of the group is necessary to achieve the prevention of large numbers of crimes even though many of those offenders imprisoned would cause no more social harm if they were set free.

We have previously suggested that one practical distinction between punishment on the grounds of dangerousness and for the purpose of general deterrence is that the attenuated link between personal guilt and the purposes of general deterrence is generally recognized. Given recognition of this link, one may expect those who sentence offenders to exercise their discretion so as not to impose deterrent punishments much harsher than would seem justified on other grounds. Where there is confidence in the capacity to predict future dangerousness and a tendency to believe that offenders predicted as dangerous will commit further crimes if not restrained, however, there is also a tendency to regard them as blameworthy (and in effect to convict them of offenses they have not yet committed) and to impose predictive sentences of imprisonment with the kind of moralistic self-confidence usually associated with excessive penal severity (Zimring and Hawkins, 1986:492–93).

Excluding Dangerousness

One point of agreement between desert theorists and the pragmatic critics of individual predictions of dangerousness was that the proper remedy for the problems they both noted was that predictions of dangerousness should be excluded from the criteria used in reaching decisions about prison sentences. From the standpoint of desert theory, the necessity for exclusion seemed to flow quite naturally and logically from the nature of the practice. Future conduct could not in fairness provide any basis for gradation in penal treatment. Though it was permissible in the assessment of blameworthiness and therefore punishment to take account of factors that were associated with future dangerousness, such as the number of current offenses and prior criminal record, these factors should be considered only insofar as they were directly relevant to that assessment rather than because they might predict recidivism and the need for incapacitation (von Hirsch, 1976:84–88).

Less obvious is the link between the more pragmatic objection to predictive sentencing and the remedy of prohibiting such predictions. If the preventive confinement of true positives is justified, and Morris never suggests any doubt about that, why then not fashion methods to curtail the incidence of false positives rather than prohibit the predictive enterprise as a whole? Two such methods would be to require very strong evidence of probable recidivism or to submit to rigorous testing the basis of individual predictions. The objective of these devices would be to reduce the error rate in predictions rather than abandoning the predictive enterprise altogether.

In the mid-1970s, however, there was little detailed consideration of such modifications largely because of the growing unpopularity of the institutions and procedures associated with the rehabilitative ideal. This may account in part for the failure of the pragmatic critics of prediction to engage in any self-conscious consideration of procedural safeguards. Thus, although Morris did not object to predictions that identified or isolated true positives, he did not count this as a positive good counterbalancing the unfairness of false positives. Under these circumstances, any possibility of even a minimal level of false-positive identification would entail the rejection of prediction, as no value was assigned to predictive success.

In the case of Morris's critique, the failure to see any virtue in the prediction of dangerousness was confined to its role in the determination of prison sentences. Morris was the first to note that security measures such as the screening systems for airport weapons detection involved wholesale predictions of dangerousness of the kind he opposed for the penal law (1984:113–114).

In later writing, when he did acknowledge the value of confining true-positive dangerous offenders, his technique for guarding against abuses of prediction involved, not creating procedures to ascertain the burden of proof, but rather confining the range of additional punishment to be imposed on the dangerous offender to amounts that the offender would be seen to deserve in any event under current community standards (Morris and Miller, 1985:37). This proposal is the focus of the next section.

Desert as a Limit on Predictive Punishment

The Morris proposal advanced in the mid-1980s, his last work on this topic as far as we know, argues that additional punishment for the dangerous offender can be justified as long as the total quantity of punishment awarded in a particular case comes within a range of punishments that is deemed appropriate for the offender according to prevailing community standards. Such additional punishments, even if exceeding the penalties imposed on other, similarly culpable but nondangerous offenders, are not problematic if they do not exceed that range.

The principle that the maximum of punishment should never exceed the punishment deserved was earlier enunciated by Morris as one of three general principles to guide the decision to imprison. He made it clear in that context that the concept of desert he employed was not that of the pure retributivist, which involves "matching" the severity of the penalty with the offender's culpability "as related to salvation or ethics," but rather one according to which the penalty is defined by what "would be seen by current mores" as appropriate. "The concept of desert" in this sense is "limited to its use as defining the maximum of punishment that the community exacts from the criminal to express the severity of the injury his crime inflicted on the community as a condition of readmitting him to society" (Morris, 1974:60, 73–74). Desert thus conceived

> is, of course, not precisely quantifiable. There is uncertainty as to the judge's role in its assessment, argument as to the extent to which he ought to reflect legislative and popular views of the gravity of the crime if they differ from his own. And further, views of the proper maximums of retributive punishments differ dramatically between countries, between cultures and subcultural groups, and in all countries over time.

Nevertheless, the deserved punishment can be defined as "what is seen by that society at that time as a deserved punishment" (Morris, 1974:75–76).

The application of this concept of desert in the jurisprudence of predictions of dangerousness

> depends on the recognition that there is a range of just punishments for a given offense; that we lack the moral calipers to say with precision of a given punishment, "That was a just punishment." All we can with precision say is "As we know our community and its values, that does not seem an unjust punishment." It therefore seems entirely proper to us, within a range of not unjust punishments, to take account of different levels of dangerousness of those to be punished; but the concept of the deserved, or rather the not undeserved, punishment properly limits the range within which utilitarian values may operate. . . .
> . . . [The] range of "not-unjust" punishments [is] measured in relation to the gravity of the offense and the offender's criminal record . . . [U]niversally these are the two leading determinants of what are seen as just punishments. (Morris and Miller, 1985:37–38)

It is apparent that there are problems involved in the application of this concept of desert as a constraint on the use of predictions of dangerousness. In the first place, it is not clear how wide the "range of 'not-unjust' punishments" might be nor what limitation on the scope of predictive judgments the desert constraint would impose. Morris's specification of desert limits provides no guidance about what sort of restriction those limits would impose, beyond the suggestion that "a Minnesota/Pennsylvania-type sentencing system . . . gives some operative and ascertainable meaning to the upper limit of desert."

In the second place, if the "range of 'not-unjust' punishments" frequently included both prison and nonprison sanctions, dangerousness might well become the sole basis for deciding between a sentence of imprisonment and its alternatives. Since desert is only a limiting principle, within those limits cases that are alike in respect of the gravity of the offense may be treated unlike on the grounds of dangerousness. This is particularly disquieting for those who insist that "persons whose criminal conduct is equally serious should be punished equally." Andrew von Hirsch makes this point:

> Suppose one treats desert as supplying only those outer limits—that the sentence must fall somewhere between *a* and *b*—and then allows the disposition to be decided within those bounds on utilitarian grounds. This would allow two offenders, whose conduct is equally reprehensible but who are considered to present differing degrees of risk, to receive differing punishments. (1985:40)

Where the difference in punishment is of a qualitative nature, as between custodial and noncustodial sanctions, the perceived inequality would be very considerable.

A third problem with Morris's use of desert as an external constraint is that perceptions of dangerousness might well influence the community's perceptions of desert. It is important to consider the question of the relationship between the community's feelings about the offender's dangerousness and societal notions of desert. In some instances, dangerousness functions not merely as an influence but as the decisive factor in desert estimates. A passage from Joel Feinberg's *Doing and Deserving* illustrates the point.

> It may seem "self-evident" to some moralists that the passionate impulsive killer, for example, deserves less suffering for his wickedness than the scheming deliberate killer; but if the question of comparative *dangerousness* is left out of mind, reasonable men not only can but will disagree in their appraisals of comparative blameworthiness, and there appears to be no rational way of resolving the issue. (1970:117)

To Bentham, in his *Specimen of a Penal Code,* it seemed self-evident that for those with "perverse anti-social dispositions . . . [t]he punishment must be more severe." His justification for this was, of course, explicitly utilitarian: the "hardened character" or the "implacable and barbarous heart" had to be "restrained by greater terrors" (1843:167). It is by no means self-evident, however, that such additional and special punishment "would be seen by current mores as undeserved" (Morris, 1974:60).

Not infrequently, special penalty ranges are advocated and provided by law because "legislative and popular views of the gravity of the crime" incorporate both estimates of the reprehensibleness of the offense and the dangerousness of the offender. There seems little doubt that not infrequently perceptions of dangerousness influence the construction of penalty scales for particular types of crime, for example, those committed by peeping toms, exhibitionists, drug offenders, and some repetitive property offenders.

This is a point of some significance when desert is defined in the terms suggested by Morris. Elsewhere in his writings (e.g., Morris, 1951), he questioned the wisdom and desirability of some protracted terms of imprisonment, but not on the grounds that they might not be deserved. Nor would it be possible for him to do so on the basis of a concept of desert which merely reflects "what is seen by that society at that time as deserved punishment" (Morris, 1974:76). Indeed, if desert is only a barometric measure of public opinion or popular prejudice, the desert limit is inherently flawed as a protection against the overuse of predictions of dangerousness.

III.   General and Specific Incapacitation

The question we are about to address has not been discussed in the modern literature on incapacitation: whether and to what extent imprisonment policies designed to promote general incapacitation suffer from the same defects as those that have provoked criticism of special incapacitation. We do not deny that many commentators have been critical of a general incapacitation strategy and that some of that criticism has raised questions about unfairness and excess (see, e.g., Messinger, 1982).

What has been missing to date is a specific analysis of the characteristics of policies of general incapacitation by those who have registered jurisprudential objections to specific incapacitation. Would Morris regard a regime that imposed one-year prison sentences on thieves rather than financial penalties as objectionable as reliance on individual predictions of dangerousness, and for the same reasons? Should he? Would von Hirsch regard such a policy as the objectionable, impermissible punishment of offenders for future rather than past crimes? Should he?

First we explore whether there are differences in kind between special and general incapacitation policies that exempt the more generalized policy from the concerns expressed about special incapacitation. Failing to find a difference in kind, we then ask whether there are differences in degree between general and special incapacitation policies that produce less concern in relation to the general policy. Our conclusion is that some differences in degree work in favor of a policy of general incapacitation to the extent that desert limits of the kind proposed in Morris's later writings are seen as a meaningful safeguard against predictive excess. If, however, an observer is worried about the capacity of incapacitative ambitions to extend the upper limits of desert to an unjustifiable degree, that problem applies to general as well as specific incapacitation.

A Qualitative Difference?

One reason those worried about the possible errors and excesses involved in individual predictions of dangerousness might relax critical scrutiny in the case of general incapacitation is that such policies entail no explicit individual predictions of dangerousness. To return to the theft example, an individual convicted of that crime is imprisoned, not because the sentencing judge thinks the defendant is dangerous but simply because he or she has been convicted of a crime for which it has been determined that the appropriate penalty is imprisonment. Since no prediction has been made about the individual's conduct, there can be no "false positive" problem, to use the language of Morris; indeed, no positive prediction at all is made. Since the imprisonment consequence flows automatically from the conviction for theft, the prison term seems to be in its totality a response to the defendant's past rather than to his or her possible future conduct and thus not to be subject to the kind of objections posed by von Hirsch.

We think that what is involved here, however, is an "optical illusion" because the implicit predictions of dangerousness that underlie classification decisions on grounds of general incapacitation can be subjected to the same objections that undermine confidence in individual predictions of dangerousness. Assume that prison rather than some alternative sanction was assigned to the theft offender category because of the belief that the incapacitative sanction would prevent, by restraint, a significant number of offenses. Is this not a wholesale prediction of dangerousness for an entire class of offenders? And will it not produce "false positives" in the same way as does individualized prediction since some of the offenders sentenced to imprisonment would not have reoffended if they had remained in the community?

If the choice between imprisonment and a lesser sanction for the group is based on the likelihood of reoffending if thieves were not confined, is the additional punishment for all thieves in effect punishment for future crime? Does it make matters better or worse in a general incapacitation policy that thieves may be imprisoned not simply because they may reoffend but because of the potential future crimes of their co-felons? A logical point is that the generality of a prediction does not remove its problematic quality. Imprisoning the thief Smith because it is felt that *thieves* may commit future crimes is not qualitatively different from imprisoning the thief Smith because it is felt that *he as an individ-*

ual is likely to commit future crimes. Indeed, to the extent that an individualized prediction may be based on the offender's own prior conduct, the individualized prediction may be less problematic than the general one. Thus, Morris in *The Future of Imprisonment* described *anamnestic* predictions based on observation of a particular individual's past behavior as less subject to error than *statistical* or *categorical* predictions derived from data relating to groups of offenders (1974:32).

The Virtues of Collective Prediction

Yet policies of general incapacitation do have advantages over selective incapacitation based on individual predictions of dangerousness in one justificatory respect. Singling out Smith from all other theft offenders as particularly dangerous and extending *his* term of confinement carries both a special stigma, the equivalent of punishing an *individual* for his or her future crimes and also invites punishment in excess of usual community standards for the individual's most recent crimes. In this situation the temptation exists to exceed the usual ceiling represented by community standards for the punishment of most of those convicted of the same crimes.

By contrast, expanding the punishment for a whole class of offenders will only succeed if the new level of punishment for the whole class is not in excess of community standards regarding just deserts for the offense. Thus, if Morris's desert limit were adequate insurance against excessive punishment based on predictions of dangerousness, the shift of decision making from individual cases to the general category would guarantee an appropriate punishment level for the category, as just defined. As long as any level of punishment below the maximum set by community standards is by definition not excessive, the issue of excess in predictive punishment has been resolved.

The problem with this solution is that notions about the dangerousness of classes of offenders may tend to increase the level of punishment desired by the community, for all of the particular class of offenders. The community wishes to punish Smith more severely than it used to feel was deserved for theft, not because it believes he is dangerous as an individual, but because it believes that thieves are dangerous as a class and as one of that class he shares that defect. The same false-positive and overprediction problems will remain; only this time they are concealed in the major premise of the general policy being pursued. Although the escalation of penalties to achieve a general incapacitative effect is a fair test of the extent to which desert limits can restrain excessive punishment, we suspect that community notions of the maximum punishment deserved upon conviction for offenses is too elastic a boundary to serve as a meaningful guarantee against penal excess.

Because of its less individualized stigma, the collective incapacitation decision may be less subject to abuse than are individual predictions of dangerousness in special incapacitation, but the question is not free from doubt. Individual predictions of dangerousness may easily lead to excessive punishment in two ways. First, the determination and attribution of dangerousness on an individual basis stigmatize in much the same way as criminal conviction. It is a small step, if a step at all, from that kind of certification to blaming the convicted offender not only for the crime that was the basis of the current conviction but for the anticipated future criminal acts on which the certification is based. The moral fervor associated with that kind of blaming can easily translate into excessive punishment. Second, the danger of unreasonably severe punishment in relation to individual predictions of dangerousness may also be exacerbated by the relatively modest pecuniary cost of excessive punishment on a retail basis. If a small percentage of a jurisdiction's prison population is singled out for additional imprisonment, this need not require any large-scale expansion of the prison enterprise.

The decision to imprison whole categories of offenders for the purpose of incapacitation may be both more costly and more resistant to stereotypes of dangerousness. The greater cost is a consequence of simple arithmetic. Escalating sanctions for larger classes of offenders not only increases the variable costs of an operating prison system but leads to the expansion of the scale of imprisonment and a considerable rise in costs in all but the most uncrowded prison systems. Moreover, the stereotype of dangerousness may be more difficult to impose on large groups of offenders. Labeling of this kind is not so readily acceptable on a wholesale basis as it is in the case of the individual determination of risk.

But large-scale stereotyping of criminal offenders can take place even when pecuniary costs are high, so the less pronounced tendency of collective incapacitation strategies to escalate punishment is an empirical hypothesis that at present is without evidential support. Furthermore, when wholesale abuses do occur, they may inflict greater harm both on offenders and on the criminal justice system than a more restricted imprisonment policy based on special determinations of dangerousness. Therefore, a shift from special to general incapacitation may carry the risk of greater harm than the excesses generated by systems based on the incapacitation of specially selected groups. These empirical questions are an important part of the modern history of imprisonment.

IV. An Ethic for General Incapacitation

Our first and most important conclusion is that more will be required to limit the punishment imposed for the purpose of general incapacitation than community values as reflected in imposed ceilings beyond which punishment is believed unfair. The problem with permitting incapacitative punishment freely up to that ceiling is twofold. First, the community's sense of retributive ceilings and floors is never precisely expressed and the gap between the minimum and maximum deserved punishment can be vast. Thus, to permit any motive for imprisonment free rein between ceiling and floor is a substantial grant of power.

The second problem with limiting incapacitative punishment only to the point at which it would exceed the community's sense of maximum deserved punishment is that public sentiments regarding the need for and efficacy of incapacitation may substantially influence the community's conception of deserved punishment. And the degree of elasticity involved may be considerable. Obviously principles and limits are needed that can inform choices even when alternative punishments are regarded as not exceeding contemporary community standards.

The one set of conditions under which we think that incapacitative punishment can operate in relative freedom from other external constraints is when the incapacitative function of the punishment is merely a byproduct of a punishment that would have been selected in any event. If we know that a year in prison would be the kind and duration of punishment selected for a particular offender independently of the prospects for incapacitation, then any incapacitative effect is a dividend the system gains at no additional cost. However, such pure byproduct cases would be relatively rare, particularly in penal systems where general incapacitation has become the dominant justification for the imposition of sentences of imprisonment.

Significantly, not all cases of imprisonment in which a nonprison sentence would be below the community's notion of the minimum deserved punishment fit into the pure byproduct category. If the public feels that nothing short of imprisonment is an appropriate punishment because of a fear of further criminality absent incapacitation, even a punishment that stands at the minimum threshold of desert may depend problematically on the expectation of incapacitation.

Assume, for example, that a majority of the public is of the opinion that sexual exhibitionists both pose a considerable danger of predatory and forcible sex crime and also have high levels of recidivism. Such a belief, if sufficiently widely held, could generate minimum deserved punishment levels that would include long prison sentences. In such cases, claims about incapacitation are a necessary part of the case for substantial imprisonment. However, if public sentiments about the need for incapacitation are not in fact justified, they should not be protected from scrutiny by the widely held public opinion that extensive periods of imprisonment are necessary and appropriate. Such cases would not fall within the scope of the byproduct rule mentioned earlier because of the role played by notions of incapacitation in the assessment of the deserved punishment.

The occasions when penal sanctions will be wholly determined without the issue of incapacitation playing some role will vary from place to place as well as over time. In the United States in the 1990s, we would suppose that the number of such cases is rather small. This, of course, is just one way of restating the dominant role of incapacitation as a justification for imprisonment, as noted in chapter 1. In most cases involving serious criminal sanctions, and in the great majority of situations that lead to imprisonment, the prospect of incapacitation is of potential impact in the choice of penal sanctions. Thus if we are correct in arguing that an emphasis on general incapacitation raises problems analogous to those presented by individual predictions of dangerousness, the problems associated with the general policy are broadly distributed throughout the criminal justice system.

The emphasis to be placed on incapacitation as a general policy for dealing with criminal offenders will affect both the type of punishment imposed on an offender and the duration and intensity of the punishment. The greater the emphasis on incapacitation, the more pronounced will be the tendency to select criminal sanctions that are designed to achieve restraint. A greater emphasis on incapacitation is thus likely to be associated with a more frequent resort to imprisonment for criminal offenders generally and for specific categories of offenders. The more pronounced the emphasis on incapacitation, the more cogent also will be the argument in favor of the imposition of longer terms of imprisonment. This may to some extent be counterbalanced when a scarcity of prison resources demands a trade-off between increasing the number of offenders incapacitated and increasing the prison terms of a smaller number of offenders.

Perhaps the most substantial impact of an increasing emphasis on general incapacitation will be on the scale of the prison enterprise itself. Whereas selective imprisonment policies have their most significant impact on the type of sanction chosen for small numbers of offenders and on the length of their prison terms, the most significant impact of an emphasis on general incapacitation concerns questions such as the appropriate size of the prison system. It is easy to see causal links between a growing emphasis on incapacitation in the late 1970s and in 1980s and the contemporaneous great expansion of imprisonment.

However, it should not necessarily be inferred that the prison boom of the 1980s was caused by the renewed emphasis on incapacitation that occurred at the same time. History teaches that very often justifications for particular penal practices are produced after their implementation and are rationalizations of change rather than causes. One counter to the notion that renewed emphasis on penal restraint caused larger numbers of prisoners is the possibility that the considerable growth in prison population was itself the cause of the voluminous rhetorical output that accompanied it. These chicken-or-egg problems are frequently encountered when we seek to unravel the relationship between theory and practice in penal history, but they are not easier to resolve because of their frequent recurrence.

Limiting Principles

When incapacitation becomes a pervasive influence on penal policy, that influence is more difficult to discern and for that reason more difficult to subject to specific controls. However, three principles help illuminate the influence of incapacitation and make it amenable to limits in legislative as well as judicial practice. First, it is desirable to maintain some notions of desert in punishment that are independent of the need for incapacitation. We do not suggest that ideas regarding the dangerousness of the typical offenders or the social value of their incapacitation play no role in the conception of the deserved punishment for specific crimes or the appropriate punishment for serious crimes generally. But there is value in separately determining what would be the appropriate minimum and maximum punishment apart from the question of restraint, to make it clear when congruence exists between punishments determined by different policy considerations and when there is significant tension between concepts of ordinal proportionality derived from retributive considerations and those that arise from a perceived need for incapacitation. If a burglar is regarded as a more significant offender than a thief on retributive grounds but the two are equal in their responsiveness to penal restraint, the contrast can only be isolated for rational debate if some separate account of desert can be constructed without reference to incapacitation.

The methodology to be used to eliminate incapacitation influences from desert ranges is a problem that deserves separate mention. Assuming that stereotypes of dangerousness inform community attitudes toward offenders, some less-than-democratic methods of isolating external influences on the assessment of desert must be attempted, possibly through the construction of a hierarchy of citizen fears or articulation of the social value of the interests threatened by different types of crime. Or it might be possible to gauge the opinions of average citizens by their responses to questionnaires about the seriousness of crimes of widely different offender types (e.g., professional criminals versus juveniles versus acquaintances). We are reasonably confident that criminology would find a number of interesting ways to deal with the problem of methodology once there was consensus on the utility of measuring desert independent of considerations of incapacitation.

A second guiding principle is that incapacitation should compete on an equal footing with other mechanisms of crime prevention. The marketplace for crime prevention includes a wide variety of wares, from environmental controls such as gun control, steering wheel locks, and cashless cabs and buses, to punitive mechanisms such as general deterrence and incapacitation. One way to pursue a balance between punitive and other mechanisms for controlling the costs of crime is to compare costs and benefits across a wide spectrum of policies. To the extent that incapacitative punishment is not required or excluded by considerations of justice, the appropriate standard of comparison is competition on a par with alternative preventive techniques.

Of course, no such competition would be possible without the careful measurement of incapacitative effects on a routine basis; and we would argue that rigorous evaluation of incapacitative imprisonment is a jurisprudential necessity. An ideological commitment to incapacitation is an inadequate basis for permitting incapacitative purposes to determine punishment policy. Yet faith rather than measurement is now, as throughout penal history, the engine for current reliance on general incapacitation in penal policy. This is both morally inappropriate and potentially wasteful of economic resources. In this sense, confidence in incapacitation as an ideology of punishment is the natural enemy of any attempt at objective determination of penal policy.

The Inadequacy of Jurisprudential Limits

Incapacitation as a justification for punishing with imprisonment is an open-ended claim. Those who wish to use prison to incapacitate, like those who hope to use it to deter, come to no natural stopping point in their claim on penal resources as long as there are criminals to punish. By contrast, some versions of retributive justification for prison use are self-limiting and punishment beyond the necessity of retribution is seen as a positive harm (von Hirsch, 1976, 1985). The amount of crime prevented by additional confinement may be modest in marginal cases, but it remains the positive result that justifies the sanction. It is not in that sense qualitatively different from other imprisonment justified by the same aim.

The limiting principles that we have just put forward are not likely to provide the definitive stopping points that an incapacitation-driven system requires. Desert limits that are determined independently of incapacitation considerations are difficult to define and very hard to sustain as a matter of practical politics. Such limits might signal when the desire to incapacitate threatens to overwhelm the other processes of criminal justice but will not of themselves effectively stem the tide. The principle of equal footing that would substitute less costly crime prevention for imprisonment would limit prison expansion for some offenses but not for all; thus it is by no means certain that investment in this form of crime prevention would lead to any net reduction of investment in prison expansion.

Limits to the open-ended justification of secure confinement are likely to be imposed by one of two different external restraints on its use. The first likely limit derives from concern about the costs, however measured, of incapacitative punishment. The costs in question may be public, public and private, solely monetary, or more broadly related to the autonomy and freedom that confinement interrupts. However, concern about the cost of additional confinement is one likely limit to incapacitation policy.

A second external limit to the use of confinement to achieve incapacitative ends may come from the limited amount of imprisonment that a society is willing to maintain. Incapacitation claims are only openended if the extent of the custodial facilities available can vary freely. If there are externally imposed limits on the scale and growth of imprisonment, incapacitation as a purpose may influence how cell space is utilized but cannot determine how much of it is available. The notion of cost concerns as a limit is discussed in chapters 7 and 8. Fixed limits on prison capacity are discussed in chapter 8.

II

RESEARCH

5

Strategies of Research

This chapter discusses three of the primary methods available for the study of the effects of imprisonment on crime rates. The first section examines the survey research of convicted offenders, the primary research method used in the 1980s, and the basis for most of the estimates of individual crime rates mentioned in the current literature. The second section discusses studies based on official records that have been obtained for samples of offenders and that have been analyzed for evidence of the effect of increased restraint on crime rates. This method was a major focus of research efforts in the 1970s, when it accounted for the majority of published research on the topic. A third section deals with the study of crime rates over time in communities where large shifts in incarceration policy have recently occurred. The identification and assessment of such "natural experiments" in incarceration policy have not been a major feature of incapacitation research to date, but should be in the future.

The essential argument of our survey is as follows. Each of the strategies of research available for the study of incapacitation has critical deficiencies. However, the problems associated with one form of research, such as surveys, are different from the problems associated with other forms, such as official-record studies. Evidence gathered from different methods can be more reliable than data generated from a single research approach. We thus argue for a balanced agenda of research, combining individual-level and community-level studies and treating offender surveys and official-record analyses of individual crime rates as cumulative contributions rather than competing approaches. A diversified portfolio of variously imperfect research methods is the most direct way of reducing our margin of error. Excluded from consideration in this chapter are models and theoretical statements relating to the volumes of crime reduction through incapacitation that are likely to be produced by particular policies.

Two preliminary points about the empirical evidence currently available seem to be beyond dispute. First, there is relatively little of it: only a few published studies and very little organized research activity on a significant scale. Second, the existing research is characterized by the clustering of particular methods at particular times. The official-record studies are creatures of the 1970s (Clarke, 1974; Greenberg, 1975; Van Dine, Conrad, and Dinitz, 1977; Petersilia and Greenwood, 1978; Blumstein and Cohen, 1979) and have not been replicated in the fifteen years since they were carried out. The surveys of prisoners conducted by the Rand Corporation were conducted in the late 1970s and early 1980s (Peterson, Greenwood, and Lavin, 1977; Peterson and Braiker, 1980; Peterson and Braiker with Polich, 1981; Marquis and Ebener, 1981; Peterson, Chaiken, Ebener, and Honig, 1982; Chaiken and Chaiken, 1982; Greenwood with Abrahamse, 1982). No new survey work was published for nearly a decade, until four were completed in the early 1990s (Dilulio, 1990; Miranne and Geerken, 1991; Horney and Marshall, 1991; and English and Mande, 1992). Indeed, one striking product of separating analysis of the new empirical research on incapacitation effects from the discussion of models is the discovery of the virtual absence of new research in the period from the early 1980s to the early 1990s. Always episodic and small-scale, research on the magnitude of incapacitation all but disappeared from most of the decade when the appetite for incapacitation fueled the largest increase in incarceration in American history.

Students of research methods in the social sciences may find that the variety of methods used to estimate incapacitation effects represents a small but interesting chapter in the history of criminological research. However, our survey is also meant to be a *consumer's* guide to methodology, of interest to those who are concerned with the substantive value of incapacitative estimates rather than with research methods for their own sake. The policy debates on incapacitation have often involved conflict about its estimated benefits, which range as we shall see, from 3 crimes per prison-year, to 14, to 187 (Peterson and Braiker, 1981:xx–xxi; Greenwood, 1982:43–47). Sometimes the divergence derives from the use of different methods or at least from the different nature of the studies, but large differences have often been the product of different interpretations of the same study. These discrepancies have made the study of research methods of interest both to the consumers of the products of this type of research and to those who produce it.

I. Survey Research

The existing research on incapacitation in the United States focuses on the estimation of the individual crime rates of classes of offenders variously located and defined. With respect to individual offense rates, one can seek information by asking persons regarded as at risk of criminal behavior about the kind and amount of crime they commit, or one can attempt to make independent assessments of the criminality of the group based on objective measures such as arrests. If one is trying to calculate the amount of crime that would have been prevented if more imprisonment had been used, one can attempt to measure the criminal activity of the target population that would have been imprisoned. But to determine the level of crime that would have occurred if a particular group had not been confined, one must either study the criminal activity of the same group at a different time in their lives to estimate what that group would have done if not confined, or one must study the behavior of persons other than those confined to approximate the crimes avoided by imprisonment in the past.

The primary documentation for estimates of incapacitation in the United States is a series of surveys of incarcerated prisoners who are asked to describe the kind and amount of crime they committed in the "window period" prior to their current incarceration. A number of such surveys were conducted by the Rand Corporation in California and the same survey instrument was also administered to prison inmates in Texas and Michigan. While these survey results are currently used and debated, the data were collected for the most part in the late 1970s. The newer survey results have not yet played an important role in the discussion of incapacitation estimates. Instead, the survey research results completed in the 1970's are not only the principal evidence on incapacitation in the American literature but they are virtually the exclusive source for estimates of individual crime rates and the volume of crimes avoided by imprisonment.

The primary data for survey estimates were produced at the Rand Corporation from 1976 through 1980 under a federally sponsored research agreement. In the first large-scale Rand survey, the subjects surveyed were 624 California prison inmates who reported the number of crimes they had committed during the three years preceding their arrest. The self-reported offenses covered nine different types of crime: homicide, assault, rape, armed robbery, burglary, theft, auto theft, cons, and drug sales (Peterson and Braiker, 1981). In Rand's second inmate survey, the sample was larger, consisting of 2,190 prison and jail inmates in three states: California, Michigan, and Texas. Jail inmates were included to provide information on offenders apparently viewed by the system as less serious. California was included for replication. Texas and Michigan were picked as representative of southern and eastern jurisdictions and because they had computerized records for sample selection (see Marquis, 1981; Chaiken and Chaiken, 1982; and Peterson et al., 1982).

Whether the exclusive reliance on prisoner surveys for incapacitation estimates is problematic depends on how reliable such surveys are as the basis for quantitative estimates of the marginal effects of increasing or decreasing prison population from current levels. For the reasons outlined later, the existing survey data are an unacceptable basis for measuring the impact of current policy and practice. The difficulties that arise from using the results of surveys of prisoners to estimate the impact of changes in imprisonment policy can be grouped under three headings:

1. accuracy of measure;
2. projection of rates to the target group of a policy;
3. uncertainty about the relationship between individual and community prevention.

Accuracy of Measure

The first set of concerns about prisoner surveys relates to whether the studies give accurate indications of what they purport to measure: the annual rates of commission of particular types of crime during the period prior to the incarceration of the survey respondents. Issues under this heading include: Do the respondents recall accurately? Do they tell the truth? Do they estimate frequencies of behavior with accuracy? How important is the phenomenon of "telescoping," in which events occurring during a period of time longer than the nominated "window period" lead to overestimates of annual rates? These are the sorts of general problems associated with surveys of self-reported criminal behavior over the course of the last half-century (see Zimring and Hawkins, 1973:321–27).

There are also special concerns associated with annualizing rates of highly active and frequently imprisoned offenders. If an active criminal is only on the streets for three months in the year prior to his current incarceration, but he committed ninety offenses in that three-month period, what should be his reported annual offense rate? Three hundred sixty? To extend the figure beyond the number of crimes actually committed is to mix hypothetical crimes with real ones in calculating the rates for imprisoned offenders. This problem is important because a small proportion of all prisoners commit a substantial share of the total crime attributable to the group. How extremely high-rate offenders are treated will have a substantial impact on the overall crime rate estimates generated by the inmate survey. Jacqueline Cohen's analysis for the 1986 panel report on career criminals has comprehensively assessed these problems (Cohen, 1986).

Overall, the sequence of surveys of inmates conducted in the late 1970s was less accomplished than the crime victim surveys of the same period in measuring and coping with accuracy problems. The Rand group had begun developing validity checks and measures, but those efforts ended with the sequence of surveys. It was not until a decade later that inmate surveys resumed on a sporadic basis (Horney and Marshall, 1991).

Appropriateness of Projections

The most important problem in estimating the incapacitation effect of current marginal changes in prison policy from the historic crime commission patterns derived from the prisoner surveys is that the historic crime commission rates should not be projected on to the target group of any contemporary change in imprisonment policy. There are three reasons why prisoner survey rates of offenses substantially overestimate the number of future offenses that any group at the margin of prison or nonprison sanctions at the current time would be likely to commit.

First, survey estimates are inaccurate on the high side during the preprison "window period" because they are not even an accurate estimate of how many offenses the same respondents would be committing, if at liberty, at the time they are interviewed in prison. To qualify for the prison sample an individual must have committed offenses during his or her preprison "window period" so that 100 percent of the sample was criminally active then. There is no reason to believe that the sample would have continued to be 100 percent criminally active if not imprisoned or if released after some confinement. Further, if a high rate of offense commission is associated with a high risk of imprisonment, the period prior to imprisonment is likely to be one in which active offenders have a higher incidence of criminal activity than at other periods in an offender's career. The technical reports of the prisoner surveys acknowledge this possibility (Peterson and Braiker, 1981; Horney and Marshall, 1991) but most commentators ignore it.

Though the use of a preprison "window period" inflates the estimate of crimes to be avoided by future imprisonment, nowhere does the available literature indicate the extent of the overestimation. The natural way to study this question is to reinterview prison survey respondents about postrelease behavior and then study postrelease arrest rates. This is not done.

A second major problem of projection is that most discussions of incapacitation generalize from the average prisoner's crime commission rate to a crime commission rate for criminal offenders just on the margin of prison sanctions. The magnitude of error introduced by this method can best be illustrated by data from the California study:

> On the average, an incoming prisoner in California committed about fourteen serious crimes per year during the three years before his prison term. However, because the average is markedly raised by the activity of a few high-rate offenders, very few offenders have crime

> rates anywhere near this average. We estimated that more than half of
> the incoming prisoners committed fewer than three crimes per year of
> street time.(Peterson and Braiker, 1981:xx–xxi)

If the relatively small percentage of prisoners who commit offenses at
very high rates are more apt to be in prison under any policy option,
their criminal proclivities will not be reflected in the crimes prevented
by modest upward or downward policy changes. Thus the marginal
incapacitation savings from the imprisonment of the close-case offender
is more likely to save three crimes rather than fourteen unless there is
no differential risk of imprisonment for high-rate offenders when
imprisonment policies are more selective.

The three-state comparison reported by Rand does not directly
address the crime commission rates of marginal offenders, but does provide good evidence that crime rates are much lower at the margin than
mean rates. The average rate of offense commission reported for Texas,
with a high rate of imprisonment, was much lower than the average rate
of offense commission for California when California's imprisonment
rates were much lower than those of Texas. The inference is that the
offense rates of those whom Texas imprisoned but California did not
were much lower than average. At least two of the most recent self-report
studies attribute lower average crime rates in their reports to high rates
of imprisonment (Miranne and Geerken, 1991) or increased rates of
incarceration (English and Mande, 1992) in the studied jurisdiction.

The third projection problem encountered when existing prisoner
surveys are used in contemporary incapacitation policy contexts is that
many of the inmate surveys are out of date by more than a decade,
including those most frequently cited in policy arguments (see, e.g.,
Abell, 1989; Gramm, 1993). Patterns of criminal activity may or may not
have changed in the last fifteen years, but criminal justice policies have
changed in ways that should reduce mean, median, and marginal
offender rates of crime.

Since 1979 the rate of imprisonment in proportion to reported
index crime has more than doubled in the United States. A major reason for the increased rate of imprisonment per 100 felony convictions is
that the states are sampling more deeply in their adult offender population when selecting groups to imprison. This should reduce the median
and mean rate of offending which a prisoner survey would find over
time in the same jurisdiction, on the same logic that predicts lower
"window period" rates of crime commission in states with higher than
average rates of imprisonment (see Cohen and Canela-Cacho, 1994:71).

One other change in offender rates is associated with the expansion in overall rates in imprisonment. The proportion of drug offenders
in the total prison population has expanded substantially since the mid-
1980s (Zimring and Hawkins, 1992). The shifting prison commitment
of drug offenders might influence the levels of crime recorded by
inmate surveys in a number of ways. Certainly, if there is any degree of

specialization among drug offenders a relative emphasis on this group will reduce offense rates based on calculations excluding drug sales and possession. The overall effect of increasing the proportion of drug offenders in the prison population is an empirical question that cannot be addressed because the inmate surveys are out of date.

The Link between Survey and Community Estimates

If an offender survey reports that twenty of its subjects committed 200 offenses last year, would locking up those twenty have reduced the level of crime in the community by 200? That is the assumption of most discussions of incapacitation. However, it fails to account for the influence of the possible reactions of persons who may have committed offenses in concert with the subject filling out the questionnaire (the so-called co-offenders), or those who may stand ready to take up any criminal opportunities the survey subject will miss because he or she is behind bars. (These group and substitution problems were discussed in chapter 3.) In either case not all of the criminal opportunities missed by the survey subject should be regarded as achieved crime prevention at the community level.

If an offender's crimes have been committed with other offenders, do those others persist without him? If the co-offenders persist, no net reduction in crime rate may occur at the community level. Without some knowledge about levels of persistence and desistence among criminal groups, the net crime reduction produced by imprisonment cannot be estimated. If the proportion of crime committed by groups is large, we know that incarcerating survey respondents will not prevent all of the crimes they would have committed had they not been incarcerated. Offender surveys can estimate the proportion of offenders convicted in groups and the offender's responses to the earlier incarceration of his or her co-offenders, but they cannot directly address the question of how large an overestimate these group processes produce.

Just as those who used to commit crimes in the company of an incarcerated offender may continue to do so in his or her absence, some of the criminal opportunities that used to be exploited by an incarcerated offender may now be pursued by other potential offenders. To the extent that the removal of an offender from the community produces *substitution*, in that others commit the crimes the incarcerated would have committed if at liberty, the number of offenses an incarcerated subject would have committed is not an accurate measure of the number of offenses prevented by his or her confinement. Because surveys of incarcerated offenders cannot address the magnitude of substitution effects, some other method of estimating incapacitation effect will need to be found to take account of this problem.

The problem of substitution in survey-based estimates of incapacitation has not been formally recognized, but the possibility of that effect may be one reason why researchers have not used the volume of acts such as drug sales reported by incarcerated offenders as a basis for estimates of reduced drug trafficking. Yet an assumption of zero substitution underlies survey-based estimates for offenses such as auto theft, larceny, and robbery, offenses for which the prospects for some levels of offense substitution seem particularly large.

Three Conclusions

Our review of the methodological problems involved in basing incapacitation policy estimates on offender survey data leads us to three preliminary conclusions. First, the margin of error associated with offender-survey-based estimates of incapacitation is large. Many of the individual problems identified above could produce large distortions in incapacitation estimates standing alone. Taking account of all the problems jointly, the best hope for using offender survey results as a policy tool is that many of these sources of error might cancel each other out rather than reinforce each other.

Second, most of the problems we have reviewed are associated with the overestimation of incapacitation savings. The problems of survey accuracy can of course produce both overestimates and underestimates of survey subject participation in crime. However, virtually all the problems associated with the appropriateness of projections of existing survey results and failures to account for group persistence and substitution tend to invite overestimation of the amount of incapacitation to be expected from marginal increments in imprisonment. Many of the problems we have encountered will compound each other rather than cancel each other out.

Our third conclusion is that many of the problems associated with the use of survey data cannot be addressed with survey methods. Careful analysis of the distribution of responses in inmate surveys can estimate the magnitude of certain projection problems (see Rolph, Chaiken, and Houchens, 1981). However, unless we obtain good estimates of the criminal propensities of the group next-most-likely to be either incapacitated if policy becomes more stringent or released if it becomes less so, an unbridgeable gap falls between what the survey can tell us and what needs to be known to estimate the impact of policies.

II. Official-Record Studies

A second group of statistical attempts to estimate incapacitation used official records of arrests for criminal activities, in some cases in conjunction with data on convictions or parole revocation, as the basis for estimates of individual crime rates. Such studies draw samples of those identified as sometime offenders from a diversity of sources, including a cohort of juvenile offenders (Clarke, 1974); a group of men arrested in the District of Columbia in 1973 (Blumstein and Cohen, 1979); a group of prisoners granted parole in 1972 (Greenberg, 1975); all those individuals convicted of felonies during a sample time period in Denver, Colorado (Petersilia and Greenwood, 1978); and arrestees in Franklin County, Ohio, during 1973 (Van Dine, Conrad and Dinitz, 1977, 1979).

The usual practice is to analyze the experience of such a group prospectively, that is, forward in time from the event that led to their inclusion in the sample, counting arrests for estimates of the prevalence and incidence of crime among members of the group while at liberty. These rates then provide the basis for estimates of the offenses that would have been avoided if the group's members had been confined as a result of previous adjudication. Thus, Blumstein and Cohen used data from the FBI computerized criminal history file relating to all those individuals arrested for homicide, rape, robbery, aggravated assault, burglary, or auto theft in Washington, D.C., during 1973, to estimate individual crime rates on the basis of the arrest histories of active offenders. They concluded that combining the individual crime rates for the different crime types, the different types of offenders committed from nine to seventeen "index" offenses per year when free (Blumstein and Cohen, 1979).

Two collaborations, Van Dine, Conrad, and Dinitz, and Petersilia and Greenwood, put forward retrospective designs in which the prior careers of persons arrested and charged with crime at a particular point in time were examined to determine how many of them would not have been at liberty to commit crime if they had been subjected to harsher penal treatment as a result of previous criminal charges. Van Dine and his associates examined the prior records of 350 offenders, convicted of homicide, robbery, aggravated assault and forcible rape in Franklin County, Ohio, in 1973, to determine the proportion that would not have been available to commit their 1973 offenses if their prior offenses had been punished more harshly (Van Dine et al., 1977). Petersilia and Greenwood searched the previous criminal records of defendants convicted of felonies in Denver, Colorado, between 1968 and 1970 to see what proportion of these persons would not have been available to commit an offense if a five-year mandatory prison term had been imposed on them for reconviction as a result of a prior felony (Petersilia and Greenwood, 1978).

This kind of retrospective analysis cannot be used to estimate individual crime rates at any point in time, because only those who are arrested for crime at the capture date for the sample are included in the study. The rate for reoffending among this group is 100 percent—an artifact of sample selection. What such an approach *can* measure is the percentage of current crime that might have been avoided if harsher penal measures had previously been imposed. The Petersilia and Greenwood study cannot tell us how many burglaries would have been avoided between mid-1968 and mid-1970 if a mandatory five-year sentence had been enforced for felony recidivists in an earlier period. However, it can provide an estimate of what percentage of adults arrested for burglary at a earlier date would not have been available for arrest under the hypothetical harsher penal policy.

Problems and Prospects

The shortcomings associated with official-record studies of individual crime rates parallel some of the difficulties identified in the discussion of survey estimates but at the same time provide a significant contrast with the tendency toward overestimation from surveys. Because the official-record studies generate individual crime rate estimates, they cannot address the problems of substitution and group criminality, noted earlier in the discussion of surveys. The offenses counted in the prospective official-record studies may include large numbers of crimes that would have been committed by other group members even if the targeted sample members had been absent. There is no way of knowing how many of the offenses committed by sample members would have been committed by others, because of substitution, if the sample group had been incapacitated.

The retrospective studies provide an estimate of prevented crime that also fails to take account of group criminality and offense substitution. In the Petersilia and Greenwood study, for example, the percentage of reduced crime fails to take into account either group persistence or substitution; the crime could therefore be smaller to the extent that either phenomenon occurred.

The significant contrast between official-record and survey studies is that official-record estimates tend to underestimate the volume of crime committed by individual sample members, and thus the potential extent of incapacitation. On the other hand, survey estimates tend toward overestimation. The specific problems encountered by official-record studies are different for the prospective and the retrospective approaches; for this reason, the two strategies should be separately discussed.

Prospective Studies

When official records are used to estimate the prevalence and incidence of criminal activity among the members of a group, inferences from these data are hampered by their incompleteness as well as the possible bias associated with official-record data. The incompleteness of the data necessitates multiplying out the rates of known offenses by some fraction to correct for unknown offenses. The possibility of bias, however, makes it extremely hazardous to do this sort of correction.

The incompleteness of official-record data as a measure of criminal activity is painfully apparent. Assume that 100 members of a sample are followed for two years, during which time 20 members of the group are arrested for a total of forty-three crimes. Does that mean that the true prevalence of crime among the group during the period was 20 percent and that the true incidence of offenses among the group was 0.43 offenses per offender? These figures must represent gross underestimates because of the compound incompleteness of official-record data on criminal involvement. Not all crimes are reported to the authorities and the very few crimes that are, are solved by processes that include the arrest of a suspect. Yet only if an offense is reported to the police and a particular individual is arrested for that offense will an attribution of the offense occur in an official-record study.

Table 5.1 provides data from the Federal Bureau of Investigation's Uniform Crime Reports on the reported clearance by arrest rates for seven index felonies. The official clearance rate varies from 14 percent to 67 percent, with high-rate offenses such as burglary and larceny being cleared by arrest less than 20 percent of the time. Further, these percentage estimates are usually regarded as grossly inflated. A burglary, for instance, is "cleared by arrest" when the police think that someone they have arrested committed a particular burglary, whether the individual is charged with that burglary or not. Restricting clearance rates to cases in which an offender is arrested for a specific crime would probably cut the clearance rate from the level shown in table 5.1 by one-half for property crimes.

Yet it is only in that small fraction of property crime for which an arrest for the specific act occurs that attribution of the offense takes place. If such attribution occurs for 10 percent of the offenses known to the police, it will constitute an even smaller fraction of all crimes occurring in the community. Under these conditions, we know that more than the 20 percent of the group with documented arrests have committed crimes in our example, and that many more than a total of 43 officially noted crimes were committed by the group. But how many more and how measured?

One natural strategy to correct for the undercount in official records is to assume the same ratio of crimes committed to arrests, for the sample

Table 5.1. Reported Clearance by Arrest Rate by Offense, 1991

Offense	Percentage
Homicide	67
Rape	52
Robbery	24
Assault	57
Burglary	14
Larceny	20
Motor Vehicle Theft	14

Source: U.S. Department of Justice, Federal Bureau of Investigation, 1992.

in a projection study, as obtained for the community generally. If 14 percent of all burglaries are cleared by arrest and 7 percent of all burglaries become the subject of a formal arrest, it is possible simply to multiply the number of burglary arrests that any individual experiences by 14 to estimate his or her crime rate. If only 7 percent of *that person's* burglaries result in arrest, this is an acceptable strategy for correcting for the incompleteness of official records. However, the "if" at the beginning of that last sentence is a very significant conjunction. If only 7 out of every 100 burglaries are the subject of an arrest, we know very little about the identities of the persons committing the other 93. They represent a "dark figure" of crime in the language of conventional criminology. A key issue is whether there is a dark figure of criminals who remain unapprehended or face lower than average apprehension risks for the crimes they commit.

One potential bias that threatens the validity of estimating crime rates by assuming that communitywide risks apply to particular group members is that these offenders may be less adept than others at avoiding detection and apprehension. It would certainly be foolhardy to assume that the members of a group selected because they were apprehended for committing a crime at any particular time committed as many offenses for which they were not apprehended as was the community average. This is because an offender has to be sufficiently inept or unlucky to get apprehended during the base time. The essential question with respect to bias is whether the unskilled or unlucky at Time 1 are apt to face higher than average apprehension chances in those subsequent periods when they are the subject of prospective studies. Caught once, are these subjects either twice as unlucky or twice as inept as others so that their personal odds of apprehension given criminal activity are above the average?

Law enforcement can take great comfort if the chances of apprehension that face samples of known criminals are no greater than average, because this would mean that there is no substantial group of active criminals who persistently run lower risks of apprehension. Some offender surveys show individuals claiming relatively high crime commission to crime apprehension ratios. These claims have been taken as evidence that known offenders run standard risks of apprehension, but the data are equivocal. Questions have been asked a relatively small number of times, and there are few ways of checking the accuracy of reports, particularly in the case of offenders who report very high volumes of offenses and have a substantial impact on the statistical aggregate in such studies. Even if accurate personal reports of risk could be obtained, it would be no easy task to determine the general community risk with which they should be compared because of the problem of unreported crime. The survey method would appear to be the best way to research the apprehension chances that known offenders face in each specific category of crime. Pending much better and more extensive data, the possibility cannot be dismissed that a group selected for their arrest experience faces elevated risks of apprehension.

Retrospective Studies

The studies that look back over the careers of apprehended offenders to determine what percentage of current offenders would have been unavailable to commit their offenses of current arrest, if harsher punishments had been imposed for previous offenses, present a different set of problems. These studies do not attempt to calculate a particular volume of prevented crime, so no multiplying of arrests or convictions needs to take place. Instead, it is assumed that the percentage of defendants who would have been off the street if they had been more harshly punished for a prior offense equals the percentage of crime reduction that could result from the harsher punishment. This technique avoids the problem of incompleteness discussed earlier but not the problem of bias. If multiple recidivism results in higher chances of apprehension for every 100 crimes offenders commit, that these offenders account for 15 percent of all burglary arrests on a particular day does not mean that they account for 15 percent of all burglaries. If the offenders' risk of apprehension is higher than for other offenders, their actual share of total crime will be lower than 15 percent.

In addition to differential risks of apprehension, retrospective studies of criminal convictions need to take account of selection biases on the part of the police, prosecutors and judges. If criminal justice agents are more likely to convict and less likely to divert recidivists on current felony charges, that recidivists account for a particular percentage of all persons convicted of felonies does not mean that they are responsible for the same percentage of felony arrests, let alone the same percentage of crime commission. Their actual percentage in these broader groups will be lower.

A further limitation of retrospective studies is that data from them cannot be translated into quantitative estimates of prevented crime. The problem is that the percentage estimates obtained from court studies are drawn from one of two distinct court systems—juvenile and criminal courts—that adjudicate large numbers of defendants. Thus when Petersilia and Greenwood estimate that a mandatory five-year sentence for offenders previously convicted of a felony would have prevented 31 percent of violent crime (Petersilia and Greenwood, 1978:609–610), the method is really limited to estimating that such a policy would eliminate 31 percent of the unknown number of offenses attributable to offenders over the juvenile age limit. All offenses committed by juveniles are not covered. A fixed percentage of an unknown quantity is, however, itself an unknown quantity.

One way of using the retrospective method to generate quantifiable estimates of total prevented crimes is to do the retrospective analysis on the basis of a sample of persons of all ages arrested for particular offenses. The percentage of arrested subjects that would have been confined under an hypothesized more stringent policy can then be multiplied by the total juvenile and adult crime volume as an estimate of the aggregate volume of crime prevented.

The main problem with this strategy is that it falsely assumes similar arrest-to-crime ratios for juveniles and adults. Because juveniles offend in groups far more frequently than adults, the same number of juveniles will typically be responsible for a smaller number of crimes than will be the case for older offenders (Zimring, 1981). A higher arrest-to-crime ratio will also result if juveniles are less adept at escaping detection. While focusing on the retrospective analysis of a total universe of arrests for a particular crime would provide a much improved context, systematic differences between adult and juvenile offending patterns would probably indicate that the adult share of arrests is smaller than the adult share of offenses committed.

A final problem involved in the use of both the prospective and retrospective studies is the substantial period of time that has elapsed since these studies were carried out. Most of the data sets discussed in this section are significantly older than even the survey data that were faulted for their antiquity in the previous section. Almost all the data discussed in this section are two decades old and some of the better data (e.g., Wolfgang et al., 1972; Clarke, 1974) are three decades old.

The great age of the official-record data sets may be less of a problem to official-record studies that are based on arrests or birth cohorts than it is for the prisoner surveys. The problem with the prisoner surveys is that the subset of the offender population that is confined to prison changed dramatically in the years since the surveys were taken, so that it is not merely time but the changes that have been documented in prison and jail populations that are a cause for concern. For birth and arrest cohorts, many changes will have taken place, but no fundamental shifts tending to undermine the comparability of samples across time have been documented. The data sets are old but not necessarily completely obsolete. One important check on the applicability of old data on birth and arrest cohorts to current conditions is replication.

III.   Community-Level Studies

The only research methods that can directly confront issues such as group processes and offense substitution are those that seek to assess the impact of policy changes on crime rates in the community. Yet if the community-level study holds the promise of being comprehensive, it is also a distressingly indirect method of estimating the level of incapacitation added or subtracted from a community setting.

The community-level assessment depends on comparing crime rates in two different areas at the same time or in the same area at different times. The former method is called cross-sectional or comparative research and the latter temporal or time series study. In a cross-sectional study two areas will differ in imprisonment policy and the analyst hopes to find out whether the policy difference accounts for any difference in crime rates, but even if the crime rates do differ, how is it possible to tell if the difference in incarceration policy has caused the difference in crime rates? Similarly, a time series study may reveal that changes in crime rate occurred after a change in penal policy. That coincidence, however, is far from proof of a causal relationship.

We call the method of community-level study indirect because logically the researcher must find a way of successfully accounting for all the other factors that might affect crime rates. Having given all the other differences between areas, or changes over time in the same area, appropriate credit for any noted changes in crime rates, the researcher can then take the residual shift in crime rates that is unexplained by other hypotheses and attribute that to the change in penal policy.

Any community-level study that comes to confident conclusions about the incapacitation effects of penal policy must first specify and measure changes other than incapacitation policy that can influence crime rates and then account for the ways that aspects of incarceration other than incapacitation can influence crime rates. Unfortunately, we lack plausible and comprehensive models of crime causation for either time series studies or area comparison so that requiring a satisfactory exposition and explanation of causal influences on crime rates is asking the researcher to attempt the impossible.

The absence of good models of crime causation makes it highly unlikely that any single temporal or cross-sectional study can produce a definitive assessment of incapacitation effects. And the primitive grasp we have of models makes us pessimistic about the value of low-budget multiple regression studies of the variety frequently found in econometrics and other social science areas. If one does not know much about what factors are responsible for crime volume, it will be very hard to compare New York's crime rate with that of North Dakota to obtain a meaningful assessment of incapacitation effects.

Yet careful and detailed studies of significant shifts in policy over time can increase knowledge about incapacitation. The kind of assessment we would favor has been called a "natural experiment" or a "policy experiment" (Campbell, 1971; Zimring, 1978). There is no mechanism in such work that serves as a definitive counter to the lack of acceptable, comprehensive models, but three tactics can be employed to minimize the dangers of mistaken causal attribution: magnitude, multiple assessments, and repetitive trials.

The best opportunity for a natural experiment will arise when large and abrupt changes in policy occur as a result of legislative or administrative change. Thus, for example, the penalty for speeding is trebled, or the probability of arrest for marijuana possession is cut by four-fifths. The size of the shift makes it more likely that the policy change will influence the crime rate and also increases the likelihood that the policy change being studied, rather than extraneous crime-related factors, is responsible for any changes in crime rate that occur. The more substantial the specific change being studied, the less intuitively likely the existence of a major confounding cause.

Also, the more abrupt a change in policy, the more plausible the assumption that the policy change rather than some other factor is responsible for any large contemporaneous fluctuation in the rate of a presumed dependant variable. Confining the time span in which we expect a major change to occur increases our confidence in causality when the hypothesized effect in fact occurs on schedule. A protracted period in which a change occurs increases the opportunity for changes in crime rates to be mistakenly attributed shifts in policy. Many more factors, other than the presumed independent variable, can have a major influence on a crime rate over a five-year period than can have an effect of similar magnitude in five months.

Another tactical response to the danger of false attribution is the use of multiple measures and what we will call qualitative tests of an hypothesis that a particular policy change is the cause of a shift in crime rates. If many factors besides the policy change can influence crime rates, one way to arbitrate between the policy change and rival hypotheses of causation is to examine closely whether patterns of crime, either demographic or geographic, have been changing in ways consistent with the causal hypothesis.

The next chapter contains a rather striking example of a cross-tabulation test of this kind that undermines the plausibility of dramatic incapacitation effects on property crime in California. We tested trends in the age of arrest for property crime on the theory that if additional incapacitation was a major explanation for the decline in that crime category, then arrests should drop more among populations subjected to the largest increases in incarceration.

This sort of qualitative test of the plausibility of causation is frequently available when social research focuses on specific changes to study them in detail over time. Persuasive qualitative tests are harder to come by in multiple regression manipulations of standard cross-sectional data sets. This is yet another reason why the policy experiment appears to be a preferable research strategy.

However, if policy experiments should be undertaken one at a time, they must be repeated frequently in a variety of different contexts before reliable estimates of policy effect can be obtained. Policy changes should be assessed in different geographical areas and at different times to diminish the chances that temporal or cross-sectional differences can successfully masquerade as policy effects. Researchers should look for contrasting policy directions. For example, the 120,000 increase in the California prison and jail population is an example of "testing up": searching for incapacitation effects when the amount of incarceration is increased. This should be balanced by "testing down," for example, by examining crime rates over time in a jurisdiction where a court order has restrained or reversed the growth of incarceration. Contrasting policy directions provide a further safeguard against false attribution.

Our discussion of policy experiment methodology has a hypothetical tone when compared to the analysis of individual crime rate research. No examples of this kind of work to date exist in the research on incapacitation. This is a significant lacuna in a balanced agenda and one that the next chapter will seek to address.

IV. Research Agenda

This section discusses both the issues and procedures of future research in incapacitation. Our analysis is based on two broad assumptions. First, a substantial effort should be made to expand, update, and improve the information available on incapacitation impacts in the American criminal justice system. The ironic pattern generated by the abandonment of research effort on the subject and the simultaneous increasing emphasis on incapacitation as a motive for penal policy needs to be reversed. Either restraint should become a much less important purpose of penal policy or efforts to measure the effects of incarceration on the incidence of crime should become a much more prominent feature of empirical research in the field of criminal justice. Only the latter change seems at all likely.

Our second broad assumption is methodological: no single research method can provide a direct and reliable answer to the core issues addressed by incapacitation research. Research that measures individual crime rates and responses to incarceration is incomplete because it cannot broach issues like group processes and substitution. Research that examines the impact of changes in penal policy at the community level is too indirect because it requires us to account for all the other things that may change in a jurisdiction over time before incapacitation can be measured as a residual. The margin of error in this kind of contortive measurement is large.

With no single definitive research strategy available, the best way to construct an accurate account of the process is to amalgamate the products of a number of different methodologies. This calls for self-conscious pluralism in the design and funding of research on this topic. An important part of the process is the avoidance of false dichotomies. The research agenda is not a matter of individual rate versus community-level studies but must include both. The individual-rate question is not a competition between survey strategies and official-record studies.

Rather it should be a combination of the greater reliability of the official-record studies with the richer detail and the motivational data that only the survey dimension can supply.

Finally, the research designs for addressing this question should not be regarded as a choice between longitudinal birth or school cohorts and samples of identified offenders only. Both methods make a unique and necessary contribution. Broad cohorts provide data on the community-wide distribution of criminal offenses, but they cannot provide the large numbers of persistent offenders that can furnish a detailed picture of incapacitation effects under various sets of systemic and social circumstances. The absence of an all-encompassing research method turns diversity of research into a necessary virtue.

What sorts of questions deserve special emphasis in the next generation of research efforts? That is an easier subject to address with respect to work on individual crime rates, given a body of research that can be scrutinized, than is the case for the relatively blank slate of community-level assessment. Thus the two categories should be separately considered.

Individual Crime Rates

In the past decade, research directed at the documentation of individual crime rates came to be thought of as the subtopic of "criminal careers" and was the subject of a National Academy of Sciences panel investigation (Blumstein et al., 1986). A modest revival has marked offender surveys in the years since the panel report but constituted a much smaller and less diverse program of research than the panel recommended.

From the perspective of incapacitation four substantive areas stand out as specially important for future research: (1) the identification of different patterns for different crime and offender types, (2) the documentation of patterns over time in individual offense frequencies, (3) the detection of variations in individual crime rates that are associated with large shifts in criminal justice policy, and (4) the study of incapacitation and specific offenses.

The Study of Differential Incapacitation

In many respects the distinction between selective and collective incapacitation is another of the false dichotomies that characterize the debate about penal restraint. All incapacitation is selective in the sense that it applies only to the persons incarcerated, and the percentage of all persons charged with crime who are incarcerated is a relatively small one (Zimring and Hawkins, 1991). The only truly collective incapacitation policy would involve locking up everybody who committed an offense.

If selectivity is the selection of candidates for incarceration from a larger pool of the eligible, it is an essential aspect of imprisonment policy at any time. There is reason to believe that focusing on particularly promising subsets of convicted felons for additional restraint will be an important pattern of emphasis in the coming years.

The enormous expansion of American prison and jails since the mid-1970s has added both to the capacity of the system to hold prisoners and to the resource burden that prisons and jails place on state and local government. If there is a cyclical character to the expansion of prison capacity, the United States may be close to the end of the most expansive part of the cycle. Without sustained general expansion in prisons, the search for policies that can enhance crime prevention through incapacitation tends to rediscover the goal of selective incapacitation, that is, a policy designed to lock up subpopulations who persistently commit large numbers of serious crimes.

Yet the research basis for selective incapacitation is dismally thin. Very little has been found in the way of offense specialization in existing studies, and persistent high-rate offenders have been difficult to identify prospectively, which is the only way that their identification could have incapacitation benefits (Greenwood and Turner, 1987). Perhaps this discouraging foundation reflects the heterogeneity of offenses and offenders in the United States. But it is also true that very little sustained attention has been paid to the detailed study of offender- and offense-specific individual crime rates. Some early work suggested that robbers and robbery might be particularly suited to preventive incapacitation (see Cohen, 1983; Greenwood, 1982). More recent work at the community level identifies burglary as the offense most responsive to general increases in imprisonment (see chapter 6, *infra*). The question of offender and offense type requires a combination of survey and official-record research on a scale many times as large as the topic received fifteen years ago. What was in effect a promising start to a research program ended prematurely when the research findings appeared to be ambiguous.

Temporal Patterns in Individual Offending Rates

In the analysis of selective incapacitation, the search for offenders with persistent high rates of crime is the equivalent of the quest for the Holy Grail. As we shall point out in chapter 8, locking up one offender who would otherwise commit 375 crimes in a year generates more crime prevention than locking up 150 offenders who would otherwise be responsible for two crimes apiece. But given persons who report extremely high rates of offending at particular times, the question arises, How fixed are their proclivities? This was identified as a key issue by the National Academy of Sciences Panel on Research on Criminal Careers in 1986 as a priority for future research:

[P]articular attention should be given to longitudinal variation during an individual's career. In what ways is the simplified model of a stable individual crime rate in error, either in terms of age trends in individual offending frequency while active or in shifts between high-rate and low-rate periods? Some researchers have identified the latter scenario as particularly common among heavy narcotics users. Are there other life events that lead to a switch into or out of a high-rate period, and does the occurrence of high-rate periods follow identifiable patterns? (Blumstein et al., 1986:205)

We agree with the priority of this topic for research and would also use it as an example of an issue that calls for samples of criminal offenders rather than birth or school cohorts as the primary subject of study. If the subject is the temporal pattern of subsequent offending for identified high-crime-rate individuals rather than the pattern of onset and desistence from criminal activity, samples from the criminal justice system would appear to be appropriate.

Policy and Temporal Changes

One conspicuous missing link in both survey and official-record studies is any attempt to compare explicitly individual crime rates associated with different groups of subjects as they may vary over time or as policies change. We observed in chapter 3 that the selective processes determining which offenders go to prison could be expected to incarcerate individuals with lower individual crime rates as prison populations expanded. That in any event is the theory, but variations by place, time, and policy have not been well documented. This type of research can be designed and executed with official records only, or with a mix of survey and official-record data.

Incapacitation and Specific Offenses

The most numerous criminal offenses in the United States are also the least important—nonaggravated thefts that are literally beyond the capacity of the justice system to count. The significant policy issues in relation to restraint concern crimes that inspire fear in the community, the so-called safety crimes: homicide, rape, and robbery as well as household burglary aggravated by violence and nonindex offenses such as arson and kidnapping. Research that does not focus on specific offenses loses track of serious but low-volume crimes. A special effort to study restraint in relation to aggravated forms of robbery, burglary and offenses of sexual violence seems required. Is the same pattern of non-specialization that holds for most offenses true of persons convicted of serious and specialized crimes of violence? What proportion of these offenses would be prevented by variations in sanctioning policy? Official-record methods seem most promising for this sort of effort.

Community-Level Studies

The large empty spaces on the map of individual crime rate records makes it difficult to talk about the studies that are required merely to fill in gaps. However, sufficient studies exist that focus on the individual offender to enable problems to be illustrated. The total absence of community-level research means that the critic is operating with an almost blank slate. For this reason our research recommendations are in effect a repetition of the general principles for community-level studies in section III of this chapter. Large variations in policy, ideally both up and down, that occur over short periods of time should be assessed using multiple measures of crime and indices of the demographic texture of offenses. A large number of different settings should be the basis for this type of research.

The most important addition that a discussion of an agenda for research can offer is an explicit judgment about the priority to be accorded to community-level studies in a balanced program of incapacitation research. The priority to be given to the assessment of natural experiments should be a very high one, at least until the problems of indirection we have discussed earlier prove to be intractable. All incapacitation initiatives are designed to have impact on crime at the community level. It would be ironic at minimum to conclude that the impact of such policies could never be measured. Meanwhile there is an oddly hypothetical quality about talking of testing the limits of community-level research methods when such approaches have not yet been tested at all. It is certainly too early to discuss detailed plans for the retirement party of a methodology that is yet to be employed.

6

Imprisonment and Crime
in California

This chapter reports materials from our ongoing study of the patterns of incarceration and crime rates in the state of California. The study is our attempt to conduct the type of "natural experiment" in penal policy advocated in chapter 5. These materials reveal the uses and limits of community-level data in the analysis of incapacitation. Section I documents the increase in prison and jail populations in California between 1980 and 1991, allowing exploration of the effects of policy change on crime rates. Section II shows the pattern of jail and prison population for the seventeen urban states in the United States during the 1980s and discusses the correlation between relative change in incarceration and crime rates over the decade. Section III discusses four different methods used to project expected crime levels in California and compares them with the crime volumes actually reported. Section IV compares projected and actual crime rates for the seven index felonies, taken as an aggregate. Section V compares projected and actual crime rates for each of the seven index felonies individually. Section VI subjects the crime reduction estimates noted in sections IV and V to a "qualitative test" to see whether shifts in age-specific crime participation are consistent with incapacitation as the major cause of declines in crime rates.

The separate strands of this study yield inconsistent support for incapacitation. Section IV shows consistent estimates that California crime would have been higher without increased imprisonment; larger crime volumes were projected for California than actually occurred. The magnitude of the gap between predicted and actual felony volume is about 3.5 felonies per extra year of confinement. Section V shows that more than 90 percent of the crime reduction occurred in burglary and larceny, while evidence for significant prevention was weak to negligible for robbery, homicide, assault, and auto theft.

But the strong indications of incapacitation effects for burglary and larceny are challenged by the pattern of arrest discussed in section VI. Rather than arrests declining most where rates of incarceration increased most markedly, among adults, the decline in both burglary and larceny was led by declines in arrest for juvenile offenders who had been subject to much less additional incarceration. This detail casts doubt on restraint as the major cause of the decline in California burglary and larceny generally.

Thus the multilevel tour of our California research results in this chapter is instructive if not fully satisfying. Though the evidence is inconsistent, the patterns revealed by the multiple layers of analysis show both the dangers of premature conclusion from nonexperimental research data and the values of multiple measurement.

To work through the separate elements of the California data is to acquire hands-on experience of the data used in this analysis and its manifold possible implications. Whatever the reader's ultimate opinion about the use of this kind of data, we believe that there is significant value in analyzing the raw material of community-level comparisons.

I.   The Growth of Incarceration in California

Figure 6.1 begins the story of incarceration in California by showing trends in the number of offenders in the state prison system between

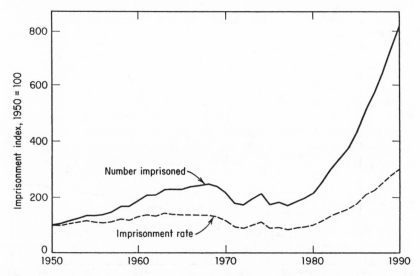

Figure 6.1. Trends in Volume and Rate of California Imprisonment, 1950–91. *Source:* California Department of Corrections Prisoners and Parolees, Series 1988–1990.

1950 and 1991 and the rate of imprisonment per 100,000 of state population for the same period. The prisoner data are end-of-year population counts from the Department of Corrections and all yearly totals are
expressed on a numerical scale that uses 1950 values as 100.

The number of prisoners climbed steadily under the pressure of
expanding population from 1950 to 1968, although the rate of imprisonment held steady. As both the numbers of prisoners and rates of
imprisonment per 100,000 population dropped in the early 1970s, with
the drop in the rate of imprisonment more than balancing increases in
state population, prison numbers did not exceed their mid-1960s levels
for fourteen years.

In January 1980, though the number of prisoners was at an average
level of 23,000 for the 1970s, the rate of imprisonment reflected in that
figure had fallen by more than 25 percent since 1968. From 1981
onward, both the numbers and the rate increased without interruption.
Even as the base rate of imprisonment increased throughout the 1980s,
the rate of increase from that base did not diminish. By the end of
1985, the prison population had more than doubled from its 1980 population base. By the end of 1990, it had quadrupled.

Figure 6.2 shows the growth for both prison and jail population in
California state and local systems by year during the 1980s. The prison
population consists of convicted felons sentenced to terms of more than
one year in state custody (though less time may be served). The jail
population consists of persons convicted of felonies or misdemeanors
who are sentenced to less than one-year terms and those awaiting trial

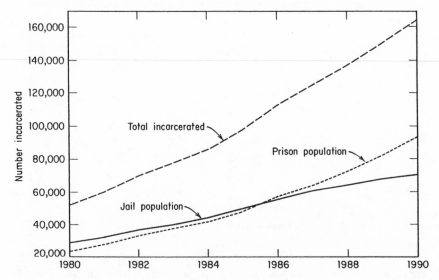

Figure 6.2. Prison and Jail Populations in California, 1980–90. *Sources:* California Department of Corrections (prison); California Board of Corrections (jail).

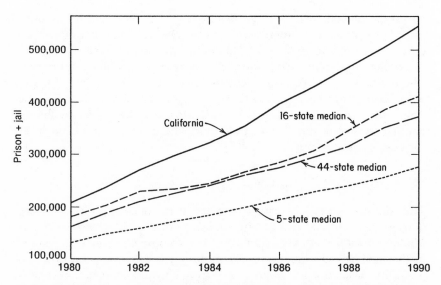

Figure 6.3. Trends in Incarceration Volume by Year: California and Comparison States. *Sources:* California Department of Corrections; U.S. Department of Justice, Bureau of Justice Statistics, 1991.

who are held in custody. The jails are under local county management in California.

Both jail and prison populations increased substantially through the decade, but the pattern for jails was slightly different. Jail population expanded from 28,946 to 70,845 over the 1980s, but three-quarters of that growth occurred prior to 1987. The annual growth in jail population had slowed to about 5 percent by decade's end.

Prisons passed jails in total population in 1985 and continued to grow in the late 1980s at a rate twice that of jails. After 1985, the California system added more prisoners each *year* than the system added in the average *decade* between 1950 and 1980. Between 1980 and 1991, California experienced seven times as much total growth in prison population as in the previous three decades combined.

The figure most comparable to English and European prison statistics is the combination of prison and jail numbers. During the 1980s, the total number in prisons and jails more than tripled from 52,000 to 170,000, while state population increased 30 percent. In one decade, this one American state added more than 100,000 extra prison and jail inmates to its correctional population.

California was not the only state in the United States to expand prison and jail numbers over the 1980s. Figure 6.3 compares annual prison and jail population movements in California with estimates of prison and jail population by year: for the other forty-nine states; for the seventeen states that contain metropolitan areas with a population in excess of 350,000; and for those five metropolitan states with the lowest rates of imprisonment increase over the 1980s. Annual jail figures are estimated using the 1983 and 1988 *Census of Local Jails* data reported for each state and linear interpolations to estimate individual jail numbers. The median rate of the five metropolitan states with the lowest prison population growth is reported here in preparation for use of these data when comparing trends over time in California to the data available on parallel trends in other states over time where large changes in incarceration did not occur.

The Search for Visible Impact

The addition of 120,000 prisoners in just over a decade is without precedent in the statistical record of imprisonment in the Western world. To what extent did that expansion affect the volume of crime experienced in the state of California? The methodology is to construct a variety of plausible projections of what crime trends would have looked like in California if no major change in policy could be anticipated and to compare the actual experience in California with different types of anticipated crime trends.

The method of estimation is both indirect and counterfactual. We have to fabricate states of California that never existed and then compare crime trends in this assortment of counterfactual domains with actual crime trends to produce estimates of incapacitation effects. But the methods we use are also inductive in an important sense. Rather than using either models or individual crime rate estimates, our primary method uses variations in volumes of crime at the aggregate statewide level, providing a basis for deriving incapacitation estimates.

The next section compares fluctuations in crime and incarceration in California with those in sixteen other states that have major metropolitan areas. This comparison is followed by a discussion of the four crime rate projection methods used to provide a basis of comparison with crime actually experienced in California. The results are then analyzed for the seven "index" offenses as an aggregate, and then for each of the seven "index" offenses individually.

II.   Cross-State Comparison over the 1980s

Before directly examining the relationship between incarceration and crime rates in California, we begin by looking at the relationship between changes in incarceration and in crime for the seventeen U.S. states that contain metropolitan areas with populations in excess of 350,000. The type of regression exercise reported here has been used as the primary tool of analysis by some social scientists but that is not our intention in these pages. Instead, the comparison of movements in crime rates and incarceration levels between 1980 and 1990 is used here to put California data in a multistate context that might help control for temporal trends over the 1980s in California that are not related to changes in incarceration policy.

We restrict this inquiry to an analysis of changes in incarceration and crime rates in the U.S. states that contain at least one metropolitan area with a population in excess of 350,000. These seventeen jurisdictions are the sample constructed for estimating crime trends in urban states over time in the section Projections of Individual Crime Volume later in this chapter. Since much crime is concentrated in large urban areas, selecting a sample of states with large metropolitan areas is superior to using all fifty U.S. states to obtain a picture of temporal trends in crime in states like California.

The question we address is whether knowing how much a state has changed its incarceration rate between 1980 and 1990 will enable us to predict the extent of a state's variation in crime rates over the same period. If states with larger than average increases in incarceration have lower than average growth in crime over the same decade this would be consistent with the hypothesis that the relationship between the greater degree of incarceration and the lesser extent of crime was causal.

There are two different ways to measure change in incarceration as a potential influence on crime. One focuses on changes in the absolute amount of incarceration while the other measures the increase in confinement as a proportion of the amount of incarceration experienced in a state as a result of previous policies. Table 6.1 illustrates the contrast between measuring absolute and proportional changes in incarceration with a hypothetical comparison between two states.

In this illustration, state A begins the period with a rate of 50 prisoners per 100,000 population which is one-third of the initial incarceration rate of state B. During the decade both jurisdictions increase their rate of incarceration by 50 prisoners per 100,000 population so that an analysis that was principally concerned with absolute levels of change in incarceration would rank the two states as exactly equivalent in performance over the decade. That is the conclusion drawn in column 3 in table 6.1. Column 4 shows that the 50-prisoner expansion represents a substantially different *proportionate* increase in the two states because of the large difference in initial incarceration rates. State A, with its lower

Table 6.1. Two Measures of Change in Incapacitation—Hypothetical Data

State	1980 Level	1990 Level	Amount of Change	Percentage of Change
A	50	100	50	100
B	150	200	50	33

base level of incarceration, doubles its incarceration rate by adding 50 prisoners, whereas state B has elevated its rate by only one-third with the same 50-prisoner expansion.

Those conducting a statistical analysis might prefer to study the influence of proportionate change in punishment policy for some purposes rather than absolute level change. Certainly a study that was trying to predict the impact of criminal justice system expansion on the system itself might want to focus on proportionate change. Perhaps also those interested in studying the effect of the threat of additional punishment as reflected in increasing incarceration rates might wish to emphasize proportionate rather than absolute levels of change. Any deterrent benefit from more stringent punitive practices should be more pronounced in state A, since by one measure the imprisonment risk has doubled in state A while merely expanding by one-third in state B. But the key variable for predicting incapacitation effects should be a change in *levels* of incarceration rather than in the magnitude of the percentage change in each state's incarcerated prison population. The behavior mechanism of incapacitation consists of restraining the criminal activity of those confined. It is therefore the number of additional confinements that should best predict the amount of crime prevented.

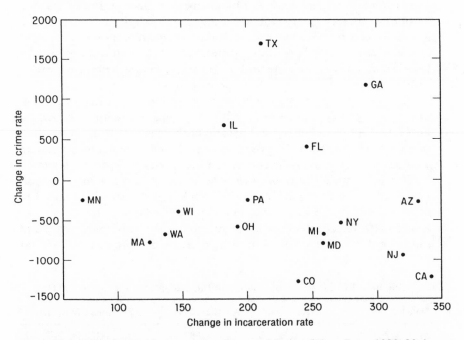

Figure 6.4. Variations in Incarceration Rate and Index Crime Rate, 1980–90, in Seventeen Metropolitan States (Correlation = –0.09). *Source:* U.S. Department of Justice, Federal Bureau of Investigation, 1980 and 1990 (crime); U.S. Department of Justice, Bureau of Justice Statistics, 1991, p. 637 (imprisonment).

Figure 6.4 displays the relation between the changes in rates of prison and jail confinement in the 1980s and changes in rates of crime. Jail populations in 1978 and 1983 are used to estimate the 1980 and 1990 jail numbers.

Each of the seventeen states is located on the scatter diagram with respect to its movement over the decade in levels of incarceration and rates per 100,000 population in levels of the seven "index" felonies used in the FBI's Uniform Crime Reports. Location on the horizontal axis shows each state's growth in prison and jail confinement. Prison confinement is measured by reported annual statistics. Jail confinement is estimated with a linear interpolation of each state's jail population in the 1988 and 1983 censuses of such facilities. All states increased levels of incarceration through the 1980s, but the rate of increase varied considerably from 72 per 100,000 in Minnesota to 342 per 100,000 in California.

The vertical axis expresses each state's change in the rate per 100,000 of the seven index felonies over the decade. The variations in crime rate noted range less widely than the changes in incarceration, with all states reporting aggregate changes between Colorado's minus 17 percent and Texas's plus 27 percent. But the visual lesson of figure 6.4 is that there is no clear patterned relationship between the amount of increased incarceration and a state's crime rate change relative to its peers. The correlation between variations in incarceration and in aggregate crime is −0.09, a minute (and statistically insignificant) negative correlation. Thus the comparison of aggregate index felonies among the seventeen states provides no basis for concluding that higher than average increases in incarceration are more closely associated with different crime outcomes than lower than average incarceration increases over the decade of the 1980s.

A number of different statistical tests can supplement the analysis reported in figure 6.4. First, we can test the impact of variations in imprisonment policy on fluctuations in the level of the four index offenses of violence. That correlation is 0.006, again suggesting no clear relation between the extent of additional confinement and the level of violent crime. The same lack of correlation (−0.09) emerges when regression results are taken for all of the forty-five states for which jail and prison data are available.

By contrast the percentage change in imprisonment rate over the 1980s in a particular state predicts the relative crime performance of a state over the same period rather nicely. The correlation between the percentage change in incarceration rate and percentage change in aggregate crime rate is −0.56, a correlation quite unlikely to occur by chance and in the hypothesized direction of larger increases in proportional incarceration being associated with less crime. The level of association suggests that variation in proportionate rates of imprisonment predicts approximately 30 percent of the variation in interstate aggregate crime rates. When the percentage of imprisonment variation is regressed against the four index offenses of violence, the statistical relation is, however, only –0.23, not a statistically significant deviation from no relationship.

Might the inverse relation between proportionate changes in imprisonment and property crime rates be evidence of an incapacitation effect after all? Certainly, theoretical explanations can be given to explain why incapacitation benefits might be larger from increasing confinement in jurisdictions with low base rates of imprisonment than would be the case when additional confinement was added where large numbers of offenders were already behind bars. But this diminishing marginal return phenomenon (see chapter 3) should not eliminate the overall relationship between increases in incarceration and rates of crime. The difference between the statistical pattern noted for percentage changes in imprisonment and changes in absolute levels of imprisonment occurs because states like Texas, Georgia, and Florida that add large numbers of additional prisoners to already burgeoning base rates of incarceration, also experience higher than average growth in crime. This sort of result is not consistent with incapacitation as a major explanation of the differential distribution of crime rates in the metropolitan states during the 1980s.

We should emphasize the limits and crudity of the comparative analysis portrayed in figure 6.4. The comparison in this analysis is between crime and incarceration at only two points in time, and the aggregate measure of crime is an amalgamation of crime categories that have not been regarded as precise measures of crime trends over time or of crime levels at any particular period. Thus it is best to regard the seventeen-state comparison as a rough beginning rather than as an exercise that could be regarded as an independent set of incapacitation findings.

III. Four Crime Rate Projections

The major problem in assessing the impact of California's prison and jail population increase lies in the determination of what levels of crime would have occurred in that state over the 1980s had the prison and jail policies not changed. No available method can provide reliable projections of future crime rates, and the exercise we must engage in is the estimation of a crime rate for a hypothetical future. Our methodology is an attempt to identify four plausible projection techniques that use different assumptions about the processes that determine crime volumes.

We use these four techniques, not in the hope of identifying a single method that is itself close to correct among the four, but rather to construct a range of estimates that will in turn produce different estimates of the crime prevention achieved by incapacitation. If the range of plausible projections includes widely varying estimates, no consistent picture of incapacitation savings will emerge. However, a lower variance between estimates may generate more consistency in the incapacitation estimates that we derive by comparing the actual pattern of California crime with crime rate projections of different kinds.

We used two strategies—cumulative and year-to-year—to project the hypothetical crime rates against which we compared the state's actual crime experience. Each strategy can be implemented either without taking into account trends in other states, or with trend adjustments. These variations produced the four separate methods of projecting crime volume in California discussed below.

Cumulative Projections

The first strategy projects crime forward from 1980 assuming stable rates of crime. Assuming that California crime per 100,000 population would stay at 1980 levels, we then adjusted estimated offense levels upward to reflect the expanding population each year from 1980 to 1991. For the aggregate and for each offense this projects a volume of crime that grows steadily through the decade to a level in 1991 that is 29 percent greater than 1980. The advantage of this method is that it is based only on California's historical and population data. The disadvantage of the method is that this projection can in no way account for changing social forces or other developments that may influence crime rates over time.

A second cumulative method was adopted to control for trends over

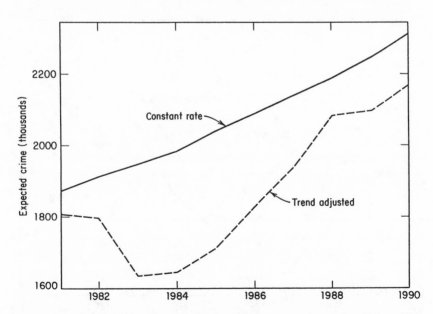

Figure 6.5. Projected California Crime Volume, Aggregate Index Felony Volume, 1981–90. *Source:* California Department of Justice.

time. The starting point for this method was again California's crime levels in 1980, adjusted each year to reflect the expanding state population.
In this model, however, the expected crime level was adjusted upward or
downward for each year to reflect the median rate change in level of the
particular crime for the median state in crime among the five metropolitan states with the lowest increases in imprisonment over the 1980s. This
strategy was selected so that the control states would reflect temporal
trends without being subject to major movements in crime as a result of
their own prison system expansion.

Figure 6.5 compares these non-trend-adjusted and trend-adjusted
estimates for the seven-crime aggregate. As figure 6.5 shows, the trend-
adjusted totals stay under the constant-rate, expanding-population
model for each year from 1981 to 1990, with the largest discrepancies
occurring from 1983 to 1986. In 1985, when the spread was greatest, the
trend-adjusted total was 15 percent under the constant-rate prediction.

Year-to-Year Projections

The two "constant level" projections use 1980 as a base year and provide
non-experience-based estimates for every year thereafter. One modification of the constant-level projection used to project crime rates in California only one year at a time was to project to the next year the current
year's crime rate with the next year's population. Since this was never

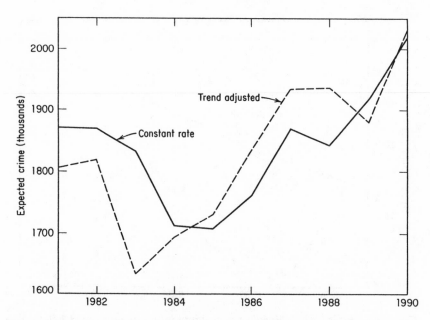

Figure 6.6. Projected One-Year California Crime Volume, Aggregate Index
Felony, 1981–90. *Source:* California Department of Justice.

Table 6.2. Characteristics of Four Projection Methods

Method	Constant Rate	Trend Adjusted
Cumulative: Single Base Point 1980		
1	Expands with population, otherwise stable	
2		Modify method 1 projection by adjusting volume by the median low imprisonment metropolitan state crime rate change.
Yearly: Previous Year Volume Is New Basepoint for Each New Projection		
3	Each year estimated as last year's crime rate adjusted for new population volume.	
4		Modify method 3 by adjusting volume by the median low imprisonment metropolitan state year-to-year change.

more than one year removed from actual data, the projected number should be much closer to actual experience than the first two methods, which hold to a 1980 base throughout. The disadvantage of this more restricted projection is that it only allows us to test shifts in incarceration over the twelve-month period that the projected figure is allowed to deviate from the actual. Within that range, the projection can be made without adjustments for trend from the other metropolitan states and also with them. Figure 6.6 compares the trend-adjusted and trend-unadjusted one-year projections for the 1980s for the aggregate of the seven index crimes.

To minimize the confounding of other state incapacitation changes, we adjusted the California crime projections by the crime trend in the state among the five lowest-increase states with the median crime rate shift as was done for the trend-adjusted cumulative projections. Though some expansion is still occurring in this bottom five in most years, the contrast between California's imprisonment trends and the comparison states' trends is substantial (see figure 6.3).

Summary of Method Elements

Table 6.2 shows the salient features of each of the four projection methods used in this section.

The purpose of the table is twofold. First, it will assist the reader by summarizing the major features of each projection method on a single reference page. Second, the table contrasts the two cumulative method-

ologies with the two year-to-year projections and illustrates trend adjust-
ment as a technique available for both cumulative and year-ahead projections. As a convention, the number assigned to each method in table
6.2 will be assigned to it for the rest of the analysis so that the cumulative constant-rate projection is method 1 throughout, the cumulative
trend-adjusted technique is method 2, and so on.

Projections of Aggregate Crime Volume

We begin our discussion of projection results with an analysis of projected volumes for all seven index felonies combined. Figure 6.7 shows
our annual projections of aggregate crime volume from 1981 to 1990.

   The cumulative unadjusted projection of method 1 has the
smoothest ascent throughout the 1980s because that is a definitional
feature of the methodology. The projected crime volume changes only
as changes in the population are multiplied by a constant 1980 rate in
each succeeding year. Method 1 also projects the highest of the four
estimated volumes from start to finish. The consistent upper-boundary
position of the method 1 estimate occurs not as a consequence of the
definition of the method but because the rate of aggregate index
offenses fell during the early 1980s and did not rebound all the way to
its 1980 per capita rate throughout the decade.

   The projected volumes for methods 2, 3, and 4 tend to exhibit
more oscillation, to stay lower, and to cluster more closely to each other

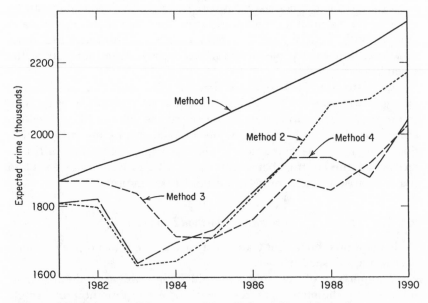

Figure 6.7. Estimates of Aggregate Crime Volume, Four Methods.

than to the method 1 projections. The greater propensity for year-to-year fluctuation is a design feature of all these methods, as is the close proximity of the two different year-to-year methods. The close agreement between method 2, the cumulative trend-adjusted projection and the two year-to-year projections is not inevitable, however, and the lower volume associated with those projections is a product of the downward trend in crime volume that characterized the early 1980s. High crime rate growth in the decade would have pushed these volume estimates ahead of method 1.

Of natural interest is the question of how much agreement and disagreement there is between the different projected crime volumes in figure 6.7. One approach to answering the question is to compute an approximate coefficient of variation among the four different series of projected crime volumes. This would measure the standard deviation between estimates, for a particular year, among the four methods divided by the mean value of the crime volume. Thus, if the standard deviation for aggregate crime as between the four methods were 200,000 and the mean value for the total of seven index crimes was 2 million, the proximate coefficient of variation would be 0.1. The lower this value, the smaller the variation between projections expressed as a percentage of the predicted crime volume. The lower the coefficient in other words, the greater the tendency for prediction methods to agree with one another.

We shall report and use this coefficient of variation measure but not without two important reservations. First, these projection methods are not independent of one another in terms of their components. All four use identical base rates for 1980 and adjust projected crime volume systematically to reflect increases in population. These and other shared characteristics will push projected volumes closer to each other than would otherwise be the case and thus reduce the estimated coefficient of variation.

Second, because of the induced closeness of the predicted crime volumes, it may well be the case that each of the projection methodologies does a better job of predicting results to be achieved by using a related method than it does of predicting what the crime volume would have been in California if no criminal justice policies had changed. There is no assurance in the close agreement between these nonindependently derived estimates that there would be a similar closeness of fit between projected and actual crime totals. Nevertheless, a relatively small approximate coefficient of variation is much more useful for those trying to use these projected volumes than a relatively large coefficient of variation, because a large coefficient of variation introduces a correspondingly large margin of known uncertainty into the projection exercise. Thus we will use the coefficient of variation between the four projection methods to compare the average disagreement noted between crime types.

Table 6.3. Coefficients of Variation in Projections of Crime Volume:
Seven Index Crimes and Aggregate

Offense	Coefficient of Variation
Larceny	0.06
Aggregate	0.07
Robbery	0.09
Assault	0.10
Vehicle Theft	0.11
Rape	0.13
Homicide	0.15
Burglary	0.18

With respect to the aggregate of the seven index felonies, the estimated coefficient of variation is 0.071, showing an average disagreement between methods equal to 7 percent of the mean total of projected crime.

Projections of Individual Crime Volume

The same procedures used to create aggregate offense projections were
used to generate volumes for each of the seven "index felonies." The
four different projections for each offense are reported in the appendix.

Table 6.3 compares values of the coefficient of variation for the
aggregate of the seven offenses and for each offense individually. The
crime-specific values are spread over a wide range, with the highest
value—burglary—being measured at three times that of the lowest—
larceny. In general, the low-volume offenses such as rape and homicide have higher coefficients of variation, whereas the two highest-vol-
ume categories—the aggregate of all offenses and larceny—have the
lowest standard error. Burglary is a spectacular exception to this pattern. The major reason for the large variation in burglary is the disagreement between the untrended cumulative estimate and the other
three methods.

IV.   Estimates of Incapacitation—Aggregate Index Offenses

How did crime experienced in California during the 1980s differ from
the projection estimates displayed in the previous section? What can
analysis of the differences between projected and experienced crime
tell us about the amount of crime prevented by the increased levels of
incarceration? We address these issues by first considering the difference between estimated and actual crime figures for the aggregate of
seven index felonies, and then discussing the patterns noted for individual offenses.

Two preliminary points about the research strategy require emphasis. First, the validity of any estimate of incapacitation using this methodology is wholly dependent on the quality of the crime rate projection discussed in the previous section. The estimates of incapacitation in this method are residuals that can only be of value if the crime projections from which they are subtracted carry meaning. Second, a variety of estimates are used in this analysis, but not as a process of elimination in which we hope to find the single best crime rate projection and therefore the most plausible incapacitation residual. Instead we hope to compare the results obtained using a variety of plausible crime projections to see whether consistency in the statistical findings can help to reduce the margin of error associated with estimates of the magnitude of incapacitation effect.

Aggregate Crime Results

Table 6.4 shows the estimated marginal reduction in crime associated with an additional prisoner-year for each of the four projection methods discussed in the previous section. Negative values, reflecting years when projections reported an expectation of less crime than California experienced, are reported with minus signs.

A number of aggregate statistical measures that can be used to describe the forty different one-year incapacitation estimates reported in table 6.4. Thirty-one of the forty estimates (or 77 percent) are positive, that is, in an arithmetic direction consistent with additional incarceration being associated with reduced crime. The median incapacitation estimate in the forty observations is between 3.2 and 3.9 per additional year of incarceration. Three of the forty observations provide estimates of marginal incapacitation greater than ten crimes avoided per year of additional confinement, and eleven observations (27 percent of the total) report incapacitation estimates over 6.

Patterned relationships exist between the type of estimation tech-

Table 6.4. Estimates of Crime Reduction by Year, Four Methods, Aggregate of Seven Index Felonies

	Method 1	Method 2	Method 3	Method 4
1981	6.4	–11.5	6.4	–11.5
1982	6.9	–3.4	7.2	1.9
1983	10.9	6.9	18.4	27.3
1984	9.8	2.0	6.1	–10.8
1985	7.2	2.7	–1.3	4.3
1986	4.5	0.8	–4.6	–5.5
1987	4.7	1.7	5.1	5.4
1988	3.9	2.4	–2.1	6.3
1989	2.9	2.3	–3.3	0.6
1990	3.1	3.2	4.1	9.0

nique used and the resulting incapacitation estimates. The untrended and cumulative estimates of method 1 produce positive incapacitation results in all ten years, whereas the two projection methods adjusted for trends in other states report 80 percent and 70 percent positive estimates. And the year-by-year analysis of method 3 provides positive estimates in six of ten observations. The cumulative and untrended estimates are also associated with higher levels of estimated incapacitation. Five of the eleven incapacitation values over 6 are reported by method 1.

The modification of California estimates to account for trends in other states produces a striking contrast between the two cumulative projection techniques. The five highest incapacitation estimates produced by method 1 (uncorrected for other state trends) occur during the period 1981 through 1985, when method 2's modification to account for rate fluctuation in the other metropolitan states produces much smaller positive estimates in three of the five years and negative values for 1981 and 1982.

The untrended estimates of method 1, though uniformly positive, shrink over the decade so that the average incapacitation estimate by this method during the first five years of the 1980s of 8.2 is more than twice the size of the average projected value of 3.8 after 1986. The estimates of methods 2 and 4 corrected for trend tend to be lower (median is 2.15) than the estimates made without comparing trends in other states (median is 4.9).

The carry-over or auto-correlation tendency in these two cumulative crime estimation techniques is quite strongly evident in the pattern of estimated results. After 1981 the sign of a previous year's incapacitation estimate predicts the sign of the current year's estimate in seventeen of the eighteen years reported in the table for the cumulative methods. For the year-to-year methods, as might be expected, the sign of a previous year's estimate predicts the current year's sign about half of the time, or in ten of the eighteen cases.

No reporting of the box score in table 6.4 would be complete without an aggregate measure of agreement across methods. For the ten years covered in the table, the four projection techniques produce unanimity in three years, 1981, 1984, and 1990, with positive incapacitation estimates ranging from 3.6 to 6.4, 0.5 to 9.8, and 1.8 to 4.8, respectively.

Table 6.5 displays the median values of incapacitation estimates from the four methods and from combining all forty estimates into a

Table 6.5. Median Incapacitation Estimate* per Person-Year, Aggregate of Seven Index Felonies

All Methods	Method 1	Method 2	Method 3	Method 4
3.5	5.5	2.15	4.6	3.1

*Median obtained as the mean of the fifth and sixth largest estimates for each individual method, and of the twentieth and twenty-first for the all method total.

single sample. The range of median estimates reflects the differences noted between the trend-adjusted and non-trend-adjusted projection methods. The median values for the two non-trend-adjusted methods were 5.5 and 4.6 offenders per person-year. The trend-adjusted totals were 2.15 for the year-to-year method and 3.1 for the cumulative total.

Though the estimates generated by the different methods varied, the range of variation was not great. There were more negative incapacitation estimates in the forty observations (9) than there were positive estimates over seven crimes per year (7). No method reported more than five of its ten annual estimates over 6, and only method 1 (with no negative estimates) had a larger number of positive estimates over 6 than of negative estimates.

One way to explore the consistency of estimates derived from this method is to test the effect of changes in the measures used in estimation, on estimated crime reduction effects. We did this in the current analysis by changing the trend-adjusted totals from the median for the five urban states exhibiting low imprisonment rise to the median for all forty-four states where the data were available. Using this broader average for trend-adjusted estimates, the ratio of positive to negative values stays at 31 to 9, but the median value of crime reduction estimated drops to 2.8 per incarceration year.[1]

For those brave enough to project a single aggregate total from the numbers in table 6.4, multiplying the 115,000 additional persons incarcerated in California during the 1980s by an estimate of 3.5 offenses per year produces a total of offenses avoided in the range of 15 percent of the 1990 crime volume.

V. Individual Offenses

Table 6.6 summarizes the findings from the separate analysis of each of the seven index offenses by each of the four previously described prediction methods. The offense-specific estimates on which the table is based are reproduced in the appendix.

Four measures are reported for each crime in the table. The first is the percentage of positive incapacitation estimates for the forty different yearly projections derived for each offense. The higher the percentage of positive incapacitation estimates, the more consistently the data produced results indicating prevention of offenses. Thus a 75 percent finding for an offense on this measure reports that three out of every four annual results are consistent with prevention through incapacitation.

The second measure reported in table 6.6 is the median value of the forty incapacitation estimates reported for an offense. This was computed by averaging the twentieth and twenty-first largest incapaci-

---

[1]The median crime reduction estimated for the cumulative trend-adjusted average substituted for method 2 was 0.7 and for the trend-adjusted year-to-year estimates 2.0.

Table 6.6. Summary of Incapacitation Measures For Specific Offenses, 1981–91

Measure	Homicide	Rape	Robbery	Assault	Burglary	Auto Theft	Larceny
Percentage of Positive Estimates	68	83	58	43	95	30	75
Median Value (mean of 20th, 21st)	.007	.055	.125	–.17	2.18	–.55	1.11
Relative Magnitude of Median	.002	.0046	.0011	.0010	.0054	.0018	.0010
Number of Years with Unanimity of Sign	1	5	1	2	8	—	4

tation estimates. The measure reported is the estimated number of crimes prevented for each additional prison-year of incarceration. Thus the 2.18 figure reported as the median value for burglary represents 2.18 burglaries avoided for each additional person incarcerated during the year. The .007 estimate for homicide signifies that fraction of a homicide avoided for each additional year of incarceration, or seven homicides avoided per 1,000 person-years of additional incarceration.

The third category reported in table 6.6 measures the *relative* magnitude of the reduction in crime reported in the data. In raw numbers the amount of burglary prevented by each additional year of imprisonment is more than two hundred times as large as the amount of homicide prevented (2.18 versus .007). However, burglaries are far more numerous in California. A relative magnitude measure attempts to control for this by relating the amount of prevention estimated to the volume of a particular offense during 1980. The median value of the prevention estimate is multiplied by 1,000 and divided by the volume of a particular offense experienced in 1990. The measure obtained is the percentage of 1990 crime that would be avoided by the incarceration of an additional 1,000 persons. As table 6.6 shows, the relative prevention magnitude predicted for homicide is in fact slightly more than one-third of the relative prevention magnitude of burglary (.002 versus .0054).

The fourth characteristic reported in table 6.6 is the number of years between 1981 and 1990 when all four projection methods agree on a sign and thus the presence or absence of prevention. This, like the percentage of positive estimates, is a measure of the consistency of the different predictions independent of their magnitude.

The index offenses portrayed in table 6.6 display three patterns. The evidence of incapacitation effects is strong for three offenses, burglary, larceny, and rape, and much weaker for robbery, homicide, and assault. For automobile theft the findings are perverse.

The statistical analysis of specific offenses provides the strongest evidence for incapacitation in the case of the offense of burglary. Ninety-five percent of the single-year estimates are positive and thus consistent with an incapacitation effect on the incidence of burglary. Every other offense has more than three times as many annual estimates that are inconsistent with incapacitation over the period studied. Burglary also has the largest estimated volume of offenses prevented: just over two per additional year of incarceration. The additional incarceration in California also appears to have had a bigger impact on the rate of this offense than on the rate of any of the others, a result indicated by the relative magnitude value of .0054, the highest for any offense. Finally, as might be predicted from the overwhelming percentage of positive estimates, the four different methods of projecting crime rates produce agreement on whether the burglary rate was reduced by incapacitation in eight of the ten years. On each of the four statistical measures the evidence favoring incapacitation is significantly stronger for burglary than for any other offense.

Larceny and rape are the two other index crimes with statistical profiles consistent with incapacitation. Larceny produces positive estimates in thirty of the forty years, and it is the only offense other than burglary for which an additional person-year of incarceration is associated with a reduction of more than one offense. Indeed the 3.29 burglaries and larcenies associated with each additional year of incarceration constitute more than 90 percent of the estimated total of crime prevented for all seven offenses. Yet the findings for larceny are weaker than those for burglary on all four of the measures reported in table 6.6. Moreover, of particular significance is the fact that the relative magnitude of the larceny prevented in less than one-fifth of the parallel finding for burglary: .0010 versus .0054. While a large number of larcenies may have been avoided as a consequence of the additional incarceration, the total savings were a much smaller faction of the large volume of larcenies reported each year in the state of California.

For rape the reduction in reported offenses for each year of additional incarceration was small (.055), but the impact of that degree of prevention on the total rate of rape was much greater than for larceny, and the percentage of annual estimates consistent with incapacitation was second to burglary among all the offenses. Thus, while the absolute numbers are small, the pattern is a consistent one.

For homicide the findings are ambiguous. Though 68 percent of all annual estimates are positive, excluding method 1 from the analysis of the homicide category produces a nearly "break-even" result of 57 percent. The median number of offenses associated with an additional year of incarceration is quite small at 7 per 1,000 and the relative magnitude that measures the percentage of all homicide avoided by additional incarceration is less than half that found for rape. Finally, the number of years in which the four different projection techniques provide the same sign for homicide is an unimpressive 1 out of 10.

Both robbery and assault produce statistical values on all four measures that are inconsistent with significant levels of incapacitation. Though 58 percent of the total number of robbery estimates are positive, fewer than half of all estimates outside method 1 are positive. And the median value of estimated incapacitation is quite small. For assault, fewer than half of the comparisons between projected and actual crime were in the direction consistent with incapacitation.

Some of the findings associated with motor vehicle theft go beyond being inconsistent with incapacitation and even suggest a reversal of the process. Seventy percent of the estimates produced when actual motor vehicle thefts are compared with projected levels are in a direction inconsistent with incapacitation. Moreover, the median crime fluctuation additional year of incarceration reported for automobile theft is the third highest in the survey, at 0.555 per additional person-year of imprisonment, but it is negative.

Does this mean that each additional year of imprisonment adds half a motor vehicle theft to California's offense profile? Any such interpretation would be bizarre. It is more likely that the models we used for projecting automobile theft were inaccurate on the low side. If that is so, it is a sobering reminder that the models that produced positive estimates for the other offenses could be erroneous on the high side. None of the findings that we have been discussing in this chapter are immune to this potential defect.

Offense Patterns

When the individual differences are grouped into categories, there is no clear pattern favoring either the high-volume offenses or property offenses in relation to the consistency of positive outcomes. The relatively low-volume offense of rape is more consistently positive on this estimate than robbery, assault, and auto theft. Two of the four violent offenses have positive estimates over 65 percent, as do two of the three nonviolent property offenses.

As might be expected, the lower-volume crimes produce lower volumes of estimated prevention even when they have consistently positive estimates of prevention. Rape has a median value of prevention less than one-twentieth of the volume of property crimes such as larceny and burglary, but as shown in table 6.6, correcting for the volume of crime estimates the relative magnitude of prevention to be higher for rape than for larceny.

One likely reason for low levels of consistency in predicted incapacitation for an offense would be that the crime projections for that offense exhibited a relatively high variation. Since estimates of incapacitation are derived from comparing actual crime experience with the dif-

ferent levels of projected crime, higher variations among crime estimates should produce more dispersion in the resulting incapacitation
estimates.

We found, however, that the coefficient of variation used in table
6.3 as a measure of disagreement between crime projections is not a
good predictor of inconsistency in incapacitation estimates for these
seven offenses. The three largest coefficients of variation in table 6.3 are
for burglary, homicide, and rape, yet burglary and rape exhibit the
highest consistency of crime rate results favoring incapacitation. Whatever explains the difference of pattern among these specific offenses,
the pattern is apparently not an artifact of differences in projected
crime rates.

Inspection of table 6.6 suggests that there is a patterned relationship between the consistency with which an offense produces estimates
in an incapacitation direction and the relative magnitude of the median
incapacitation estimate for that offense. The two offenses that exhibit
the most consistency of incapacitation—rape and burglary—also have
the two highest magnitudes of incapacitation effect, as measured by the
size of the median incapacitation effect as a fraction of the crime volume of an offense in 1990. The only apparent exception to this pattern
is larceny, with a consistency estimate of 75 percent but a relative magnitude of incapacitation estimate at the bottom of the distribution for
crimes with indications of consistent incapacitative effects.

The Problem of Unreported Crime

One difficulty with using the estimates from this analysis as a veridical
description of the community impact of California's increased incarceration is that our study only measured trends in crime reported to and
by the police. By some estimates one-half of all offenses that citizens
report in telephone surveys do not end up as crimes known to the
police and are thus excluded from the categories that form the basis of
our estimates in tables 6.5 and 6.6. While those unreported crimes are,
with some exceptions, mostly found in the least serious crime categories
and are usually among the less serious forms of those offense categories, the failure to measure trends in offenses not reported to the
police is a definite shortcoming in the method we use. It also represents
a problem with no easy solution.

If only 50 percent of all offenses are reported to the police, why
shouldn't we simply double the estimates of incapacitation on known
crimes to estimate the community impact of additional incarceration?
Doubling the 3.5 index offenses in table 6.5 would produce an annual
prevention estimate of seven felonies closer to the survey-based results
discussed in chapter 5. The problems involved in coming to any such
conclusion are manifold. First, there is no clear indication that trends in
unreported offenses follow trends in reported offenses. Yet doubling the

identified decline in reported crime assumes that trends in reported crime are an accurate index of fluctuations in total crime.

Second, the doubling measure would assume that offenses not reported to the police were distributed across the population of potential offenders in the same pattern as offenses that *are* reported to the police. However, victims are less likely to report offenses committed by family members and intimate acquaintances, unsuccessful attempts and behavior that they think the police might not regard as seriously criminal. Thus there are a number of reasons to suppose that unreported offenses may be distributed across a broader cross section of the general population and are less concentrated in the population of known offenders than are the crimes that are known to the police that form the basis for the estimates in tables 6.5 and 6.6.

We have no doubt that the estimates in tables 6.5 and 6.6 do not include some offenses that would have been committed in California except for the increase in incarceration, but that would not have been reported if they *had* been committed. Yet there is no way of estimating either the quantity of such offenses or its relationship to variations in reported crime.

Rather than artificially inflating the estimates we have developed, a more appropriate way of putting into context the crime prevention achieved by additional incarceration is by expressing the prevented offenses as a percentage of all known reported offenses—in essence, comparing apples with apples. At two burglaries per incarcerated year, the estimate in table 6.6 is more than five times as much crime prevented relative to 1990 volume as is the one theft per additional year.

VI.   A Qualitative Test of Incapacitation as Cause

As will soon be evident, the preceding analysis stops well short of proof that the increasing rates of incarceration in California were the cause of decreases in crime volume noted for burglary, larceny, and rape. In this section we will report an attempt to push the analysis one step further by examining whether changes in patterns of arrest are consistent with incapacitation as the major cause of lower crime rates. The focus of our initial analysis is on burglary, the index offense that alone accounts for more than half of all incapacitation estimated in the preceding section and the crime category that was most consistent with incapacitation on all of the measures displayed in table 6.6.

If incapacitation is the major explanation for decreasing rates of burglary, the rate of burglary should be reduced more substantially among those population groups for which incarceration increased the most. The first step in the analysis is to contrast trends in levels of incarceration for offenders under 18 and over 18 in the 1980s. Table 6.7

Table 6.7. Trends in Secure Confinement in California For Juveniles and Adults

Offenders	1980*	1990*	1980–90 Percent Change
Juveniles	8,321	11,632	+40
Adults	51,578	158,143	+207

*December 31 populations for 1979 and 1989, respectively, for juvenile halls, county camps, California Youth Authority Facilities, jails, and California Departmental Corrections.

Sources: California Department of Corrections; California Department of Justice; California Youth Authority.

compares changes in incarceration levels for offenders under 18 with parallel data for adult offenders.

Though the incarceration of both age groups increased during the 1980s, the rate of increase was more than five times as great for adult offenders. Adjustments for change in population still leave a greater than 3 to 1 advantage for adult incarceration.[2] Did participation in burglary diminish more among the group that experienced a larger measure of additional incarceration? We can approach this question if we can find measures of criminal participation by separate age groups.

There is a no direct statistical measure of participation in burglary to test this hypothesis, but trends in arrests should be a respectable index of burglary participation for various age groups. We should expect to find that the larger increase in incarceration among adults would reduce the proportion of total arrests attributable to adults as it prevented a larger number of adult burglaries. Further, this shift away from arrest concentration among adults should be largest for those offense categories when more substantial amounts of incapacitation took place. Table 6.8 contrasts 1980 and 1990 arrest profiles by age for the burglary category that displayed an impressive appearance of incapacitation in table 6.6 and for robbery, for which the evidence of incapacitation was slight.

The pattern displayed in table 6.8 is exactly the opposite of that which would be most consistent with incapacitation as the major cause for the decline in burglary. Since the largest increase in incarceration was among adults, the relative participation of adults in burglary should be reduced over the course of decade. Instead, adult burglary arrests increased 19 percent, whereas juvenile burglary arrests decreased 36 percent. If arrest is a consistent indicator of trends in criminal participation over time, it is the adults who are locked up in California and the juveniles who stop committing the burglary.

[2]Rates of incarceration per 100,000 for ages 14 to 17 increased 46 percent assuming a constant percentage of subjects that old in the youth corrections system; rates per adult ages 18–35 increased about 148 percent. (See tables 6.7 and 6.10.)

Table 6.8. Trends in Burglary and Robbery Arrests,
Juvenile and Adult, 1980 and 1990

Arrests	1980	1990	1980–90 Percent Change
Burglary			
Juveniles	36,814	23,745	−36
Adult	47,346	56,166	+19
Robbery			
Juvenile	7,354	7,780	+6
Adult	19,361	24,264	+25

Source: California Department of Justice.

are analyzed. Since the burglary rate was far below its expected levels
while the robbery rate was not, the concentration of adult offenders
should dissipate more for burglary than for robbery. However, robbery
arrests increased 6 percent for juveniles in the 1980s, whereas they
increased 25 percent for adults. Thus the contrast between juvenile and
adult arrest patterns for robbery is less pronounced than for burglary.
The data on robbery arrests also make it less likely that a major change
in arrest reporting and classification has produced the distinctive pattern shown for burglary.

Table 6.9 expands the analysis to show arrest trends in the five
index felonies other than burglary and robbery. The pattern in the
table is quite clear. Larceny and rape, the other candidates for significant incapacitation in the analysis in section V, both exhibit the same
pattern as burglary—a decrease in juvenile arrests but an increase in

Table 6.9. Arrests for Five Index Crimes, California, 1980 and 1990

Index Crime	1980	1990	1980–90 Percent Change
Larceny			
Juvenile	15,233	11,154	−27
Adult	35,814	55,931	+56
Rape			
Juvenile	672	630	−6
Adult	3,836	4,218	+10
Homicide			
Juvenile	445	658	+48
Adult	2,778	3,224	+16
Assault			
Juvenile	8,469	11,379	+34
Adult	40,486	95,402	+136
Auto Theft			
Juvenile	11,072	17,101	+54
Adult	18,442	30,120	+63

Source: California Department of Justice.

Table 6.10. California Population, by Age

Age	1980	1990	1980–90 Percent Change
14–17	1,601,816	1,523,439	−5
18–35	7,829,332	9,666,973	+23

Source: U.S. Department of Commerce, Bureau of Census.

adult arrests. If these arrest trends measure each group's participation in each crime category, whatever is reducing involvement in larceny and rape is more effective among juveniles than among adults. But according to table 6.7, the incarceration increase was much higher among adults.

The three crimes reported in table 6.9 that did not show strong evidence of decline from expected levels also fail to show the characteristic pattern of juvenile arrests declining while adult arrests increase. Whatever generates this pattern seems responsible for the lower-than-expected rates of burglary, robbery, and rape.

Information on demographic trends in California clarifies but does not explain the age-specific patterns noted in arrest statistics. Table 6.10 shows changes in population volume for the age groups 14–17, which account for the overwhelming majority of index arrests among juveniles and for the crime-prone years of young adulthood. The contrast is sharp, with the large expansion in young adults juxtaposed to a 5 percent decline in older juveniles. However, incapacitation increases normed to population still favor the adults by more than 3 to 1, and the benefit of the decline in juvenile population only occurs for burglary, larceny, and rape arrests. Thus the specific mystery of these offenses is unresolved by the demographic shifts.

It is difficult to know how much emphasis should be put on the paradoxical pattern revealed in tables 6.8 and 6.9. Incapacitation as a significant explanation for diminished burglary and larceny is by no means impossible, given the trends in age-specific arrest. The fraction of high-rate offenders may be higher among the 3,300 additional juveniles confined than among the 106,000 extra adults, but if extra juvenile confinement is the mechanism for reducing reported burglary in California, it is close competition for the miracle of the loaves and the fishes. If extra adult incarceration is the cause, the arrest trends we discovered are inexplicable. At minimum, then, these data should inspire substantial caution about the role of additional incarceration as the major explanation for decreasing offenses in California in the 1980s. We believe that a convincing explanation of declines in burglary and larceny must give more credit to trends among juvenile offenders than it is plausible to extract from incapacitation as a leading cause of the decline.

Conclusion

The major undertaking of this chapter, a series of crime rate projections against which actual crime experience in California was compared, provided consistent indications of a reduction in reported "index" felonies of just over three per year, with subsequent analysis showing that the reductions were concentrated in burglary and larceny categories. A third offense, rape, showed indications of reductions from expected levels but with lower volumes. There were no indications of substantial incapacitation benefits for homicide, assault, and robbery, and auto theft rates were much higher than expected.

To confirm the dominant role of incapacitation in the lower rates of burglary, larceny, and rape, we compared data on extra incapacitation with information about arrest rates by age. The extra incapacitation was concentrated on offenders over 18. However, the reduction in rates of burglary and larceny appeared to be concentrated on offenders under 18 because juvenile arrests for these crimes went down while arrests of older offenders for these offenses increased. This is the opposite of the pattern we would expect if the additional incapacitation were the major cause of the crime reduction.

So the evidence that additional restraint was the primary cause of the noted crime reductions is much more equivocal than first appeared. There are no other obvious causes, such as a big drop in populations at risk, or increased deterrence, that would fit the peculiar pattern of fewer juveniles–more adults noted in tables 6.8 and 6.9 for the three crimes for which significant incapacitation seemed present. Just because these data do not fit the incapacitation hypothesis very well does not, however, mean that they *do* fit a plausible rival theory. A parsimonious theory to explain the texture of the decline is not currently at hand.

Further, there is no way to adjust quantitatively the estimates from sections IV and V to take account of the counter-intuitive pattern of age-specific arrests. We called the cross-tabulation of arrest trends by age a "qualitative test" because rather than modify the magnitude of particular incapacitation estimates, it reinforces or undermines the conclusion that the noted reductions were caused by additional restraint. The estimates in sections IV and V remain unchanged but are merely less likely to have been the result of the imprisonment policy changes that were occurring.

If the evidence from this analysis is equivocal, then what should happen next? The mystery surrounding the decline in California crime should provoke more detailed examination of California data and meticulous natural history studies in a number of other jurisdictions with dramatic changes in imprisonment policy. Such studies should include examination of both rapid increases in incarceration in the California style and case studies in which increases were halted or reversed by court decree or other circumstances. A series of carefully documented individual case histories is the best hope for finding valid incapacitation estimates.

Among the California refinements that should be attempted are a detailed examination of burglary rates and reporting, and a series of "geographic cross-tabulations" to examine whether the communities that increased incarceration more in California enjoyed disproportionate subsequent declines in the offenses that seem most responsive to statewide increases in imprisonment.

Why suggest these refinements of a research strategy that produced such equivocal results? There are two reasons. First, the weakness of alternative methods still requires a major investment in the natural experiments of real change in imprisonment policy. Second, the methods used here are not weak: the equivocal results found show only that the evidence available in California is not as conclusive as might first appear.

The problems with other methods of estimating incapacitation have already been rehearsed. Individual rates are grossly overstated in prisoner surveys, and the community impact of imprisonment cannot be addressed by measuring individual crime rates. Cross-sectional multiple regression studies are at the mercy of suspiciously incomplete models of crime causation and manifold problems of measurement. Aggregate time series analysis at the national level seems ludicrous. With individual states as units of analysis, it becomes the same basic strategy as the natural history, but with much less detail in data. Certainly, most cross-sectional regression studies would not produce the impasse described in the previous two sections, but that is a weakness of multiple regression.

The methods used in this analysis of California appear to be coherent and sensitive. One should not confuse the strength of a research method with the strength of the evidence available for a particular study site. If the California arrest patterns had been consistent with larger crime reductions in the most incarcerated groups, much confidence could be invested in the conclusion that incapacitation played a major role in the declining rate of crime. That the methods used detected an inconsistent pattern is part of the strength and sensitivity of the overall design. If the empirical correlates of incapacitation are more consistently present in other settings, careful multiple-measure assessment over time will discover consistent trends. If the underlying evidence for a relationship is equivocal, then the best methods of research will help reveal the full dimensions of the inconsistent evidence.

III

POLICY

7

Of Cost and Benefit

No adequate computation of what crime costs in terms of current exchange can be made or even imagined. . . . After all, it is impossible to avoid a painful feeling that any measurement of crime by monetary standards is a belittling of the subject . . . .

Eugene Smith
The Cost of Crime (1901)

A considerable literature on the cost of crime is already extant. Much of this literature is wholly fallacious. Most of it is misleading in its general implications. Only a very small part of it is written with any conception of the methodological difficulties inherent in the problem. It is strange that so many pages of printed matter should have been produced with such a small expenditure of reflective thought.

E. R. Hawkins and Willard Waller
Critical Notes on the Cost of Crime (1936)

I. Issues of Cost and the Question of Incapacitation

This chapter discusses attempts to measure the public and private costs of crime in dollar amounts and to translate those cost estimates into dollar sums that represent the monetary value of crimes prevented as a result of incapacitation. There is no logical reason why a study of incapacitation should have to analyze the monetary cost of crime. It is not difficult to imagine a sustained dialogue about the incapacitation effects of imprisonment in which crime prevention benefits were not converted into monetary values. Indeed no extensive monetization associated with the discussion of incapacitation effects existed until quite recently.

Yet dollar cost and dollar benefit have become the dominant terms of reference in the discussion of incapacitation effects in recent years, and the assumptions and conclusions of many current studies cannot be comprehended without extensive reference to monetary estimates and the ways in which they were derived. In this chapter we will first discuss the attraction of approaches that provide money cost estimates for crime and crime prevention. We will then examine two "cost of crime" studies in some detail. The third section of the chapter will survey theoretical material that supports different strategies for assessing the costs of crime and the costs of crime countermeasures. The concluding section will consider constructive uses as well as the considerable limitations of monetary cost estimates in the discussion of criminal justice policies.

Our intellectual ambitions for this chapter aim toward the deconstruction of the cost-benefit slogans currently used in the discussion of incapacitation. Dollar cost estimates provide a useful means of comparing different types of crime prevention with each other and can be helpful in the study of offenses where most of the harm caused by the offense is economic, but the wholesale translation of heterogeneous criminal behaviors into aggregate dollar volumes is usually misleading and always inadequately supported by meaningful conceptual foundations.

Financial Attraction

Why the sudden popularity of measuring both crime and crime countermeasures in dollar terms? The attraction of the monetization of crime and crime countermeasures is in fact twofold. First, translating all of the different types of harm associated with crime into dollar terms allows the comparison of different types of crime on a single standard. Robbery, theft, and assault are different kinds of behavior, and finding a single standard by which the harms associated with each can be reliably measured would permit selective judgment about which offenses should command a larger measure of public concern and public resources.

Second, a single, dollar standard for the costs of crime allows us to measure both the crime and the crime countermeasure in the same pecuniary units. If a crime prevention strategy—police patrol, imprisonment, or burglar alarms—can be expressed in terms of cost, and the offenses it is designed to prevent can also be expressed in monetary values, it should be possible to arrive at definitive judgments about whether a particular countermeasure produces benefits worth its cost. If a particular type of theft costs $X per occurrence and a unit of policing designed to prevent it costs $3X, any crime prevention total greater than three per unit is cost justified, and any crime prevention product smaller than 3 would suggest that the countermeasure is not justified. Monetization of this kind turns policy analysis into simple arithmetic.

The major disadvantage associated with the monetization of crime and crime prevention is that it generates an illusion of commensurability when the attributes being factored into a money cost standard are in fact incommensurable. A single pecuniary standard for the consequences of crime promises a scheme in which homicide and rape are assigned dollar values that allow them to be scaled in both cardinal and ordinal dimensions along with automobile theft and larceny. A monetization schema also facilitates the direct comparison of the dollar costs of rape with the dollar costs of an antirape police patrol.

The standard by which monetized estimates and strategies should be judged is the extent to which their use justifies a claim of single-standard commensurability. Monetized values of this sort are rarely neutral in impact. Either they greatly advance our capacity to do policy analysis when they successfully function as a single standard, or they diminish our capacity to make intelligent policy choices by generating an illusory standard that invites erroneous judgments. It is precisely the same features of the schema that provide both its advantages and its drawbacks.

Under these circumstances the key question is, How far short of their unifying ambitions are the current achievements of monetization schemes? To answer that question we will first examine two studies of cost from the current criminal justice literature and then compare those studies with the available theoretical work on the economics of crime.

The selection of papers for review was not difficult. The two to be discussed here have been the most frequently mentioned studies in academic and policy discussion of cost analysis. The paper by Cohen (1988) is the first major treatment of the matter in academic literature. The Zedlewski analysis (1987) is the most frequently cited work in policy discussion of the subject. Each would be included by consensus in any short list of relevant publications.

II. Two Studies

Victim Costs (Cohen 1988)

Mark A. Cohen wrote "Pain, Suffering, and Jury Awards: A Study of the Cost of Crime to Victims," which appeared in *Law and Society Review* in 1988, with two main purposes in view: to provide estimates of the cost to victims of individual crimes and to demonstrate the policy implications of those estimates. The bulk of the paper deals with the cost of crime to victims; the possible policy applications of victim cost estimates are dealt with more briefly by means of examples.

Cohen's study is both original and ingenious. It is original because previous studies of the cost of crime to victims have largely ignored the pain, suffering, and fear endured by crime victims and concentrated on the out-of-pocket costs of crime to victims, such as stolen or lost property, medical costs, and lost wages. Cohen's is the first to estimate the monetary cost of pain, suffering, and fear caused by individual crimes. His study is ingenious in the method used to arrive at crime-specific estimates of the pain, suffering, and fear components of crime.

Cohen begins by examining jury award data to arrive at these estimates. The conversion into dollar amounts is achieved by examining court awards in personal injury cases for similar injuries and on estimates of the value of life. He then derives an estimate of the total cost of crime to victims by combining information on the distribution of injuries and deaths with estimates of the monetary value of the pain, suffering, fear, and death.

Having arrived at estimates of the three main components of the cost of crime to victims—that is, (1) direct monetary or out-of-pocket costs; (2) pain, suffering, and fear of injury; and (3) risk of death—Cohen then combines estimates of these three costs to arrive at an overall victim cost for each type of crime as shown in table 7.1.

The first column of the table reports the direct or out-of-pocket costs incurred by crime victims, including any property or theft losses, medical and psychological counseling expenses, and time lost from work. The second column reports the pain and suffering estimates that were calculated as simply the fraction of victims who incur each type of injury, multiplied by the respective pain and suffering estimate. The third column estimates the monetary value of the risk of death associated with each crime, the probability of death being derived by dividing the number of murders associated with each type of crime by the number of those crimes committed, which is then multiplied by the estimated value of a statistical life of $2 million. In the last column of the table estimates of the three types of cost are combined to arrive at an overall victim cost for each type of crime.

Cohen then gives examples of policy applications, using victim cost

Table 7.1. Average Cost of Crime to Victims

Crime	Direct Losses	Pain and Suffering	Risk of Death	Total
Kidnapping	$1,872	$15,797	$92,800	$110,469
Bombing	24,737	7,586	44,800	77,123
Rape	4,617	43,561	2,880	51,058
Arson	14,776	6,393	12,380	33,549
Bank robbery	4,422	10,688	3,700	18,810
Robbery	1,114	7,459	4,021	12,594
Assault	422	4,921	6,685	12,028
Car theft	3,069	—	58	3,127
Burglary	939	317	116	1,372
Larceny				
Personal	179	—	2	181
Household	173	—	—	173

Source: Cohen, 1988, p. 546.

estimates. The most striking example deals with James Austin's 1986 study of the effect of the Illinois early release program on both the reduction of prison population and consequent increase in crime. Austin's study concludes that:

> Between 1980 and 1983, the Illinois Department of Corrections made an early release of over 21,000 inmates in response to a prison crowding crisis. During this period, over 5,900 prison years were averted and the projected prison population was reduced by approximately 10 percent . . . Also, early release substantially accelerated the amount of crime suffered by public, but contributed to less than 1 percent of all crimes reported in Illinois . . . However, overall early release proved to be cost-effective. (1986:404)

Cohen points out that Austin did not have estimates of the pain and suffering endured by crime victims but used Bureau of Justice Statistics estimates of the out-of-pocket costs. He goes on to say that if one were to use the estimates shown in table 7.1 to recalculate the cost of the early-release program the opposite conclusion would follow. Austin had estimated the program's benefits to be about $49 million (a figure due solely to averted prison costs) and the program's costs (including the additional criminal justice costs associated with supervising the early releases and reprocessing recidivists into the system, as well as victim costs) to be $17 million. However, Cohen explains that including the cost of victims' pain and suffering increases the cost of the program to about $110 million. Austin had estimated that stated in terms of the cost "per early-release inmate" the program resulted in a *saving* of $1,480, whereas the inclusion of pain and suffering changes that saving into a *cost* of $2,870.

Cohen also gives a number of other examples of the potential use of the victim cost estimates. Thus, using the data relating to crime reduction as a result of increasing sentences for robbers from a 1983 article by Jacqueline Cohen (pp.70–73), he maintains that through the use of crime-specific estimates it can *now* be calculated that increasing the average sentence length for robbers from forty-two to fifty months, based on the number of robberies and inmates and the estimated cost to victims of $12,594 per robbery, would have a benefit-cost ratio of about 3 to 1.

Another example relates to determining whether the deterrent effect of increased sentence lengths is worth the cost of additional prison capacity. Cohen points out that Edwin Zedlewski had concluded that, in general, the results of a benefit-cost analysis "overwhelmingly support the case for more prison capacity" (Zedlewski, 1985:6), but he notes that Zedlewski's analysis was hampered by the lack of crime-specific estimates and was thus limited to recommendations concerning the average length of sentence for all crimes in general.

Cohen develops his argument by reference to Donald Lewis's 1986 article "The General Deterrent Effect of Longer Sentences" (Lewis, 1986:47–62). Lewis had surveyed forty-nine studies involving fifty-two data sets on the deterrent effect of longer sentences and estimated the mean "elasticity of severity," which he defined as "the percentage change in the crime rate for a specific crime divided by the percentage change in the average sentence served by prisoners convicted of that crime, other things being equal" (Lewis 1986:48). Cohen uses Lewis's estimates of the elasticity of severity to conduct a benefit-cost analysis of increased prison sentences for each crime on the FBI index, as illustrated in table 7.2. The text accompanying the table runs as follows:

> Assuming a 10 percent increase in time served, one can calculate the reduction in each type of crime. For example, a 10 percent increase in time served for rape would result in about 5,500 fewer rapes. Based on an estimated cost to victims of $51,058 [from Table 7.2], this would yield a benefit of $282 million. Of course, each convicted rapist would now spend 10 percent more time in person, which would increase the average time from 54.3 months to about 60 months. Based on the number of incarcerated rapists and the cost of imprisonment, this results in an added prison cost of $104 million, or a benefit ratio of 2.7. (Cohen, 1988:551)

Cohen concludes that the analysis suggests that longer prison sentences are warranted for rape, assault, and auto theft and that shorter sentences might be warranted for burglary and larceny. He adds, however, that these are not policy recommendations, as the estimates in

Table 7.2. Benefits and Costs of 10 Percent Longer Prison Sentences

Benefits and Costs	Rape	Robbery	Assault	Burglary	Larceny	Auto Theft
Benefits						
Number of crimes	78,920	506,570	653,290	3,129,900	6,712,800	1,007,900
× percentage of reduction	7.0	4.7	6.0	3.4	2.8	2.4
= number of reduction	5,524	23,809	39,197	106,417	187,958	24,190
× cost per crime	$51,058	$12,594	$12,028	$1,372	$180	$3,127
= total benefits (millions)	$282	$300	$471	$146	$34	$76
Costs						
Number of prisoners	9,611	46,259	25,776	78,372	44,791	8,031
× average months in prison	54.3	36.3	28.5	19.4	17.3	16.3
× 200						
= total costs (millions)	$104	$336	$147	$304	$146	$28
Benefit-cost ratio	2.7	0.9	3.2	0.5	0.2	2.7

Source: Cohen, 1988, p. 552.

Table 7.3. Comparison of Monetary Harm to Sentence Length
for Bank Robbery

Offense Characteristics	Medial Months Served under Sentencing Guidelines	Percentage of Marginal Increase in Sentence	Monetary Estimate of Harm to Victim	Percentage of Marginal Increase in Harm
Bank robbery without weapon	37.0	0	$6,282	0
Possession of weapon	58.5	58	$21,100	236
Bodily injury (weapon present)	70.5	20	$25,586	21
Serious bodily injury (weapon present)	87.5	24	$54,535	113
Permanent or life-threatening bodily injury (weapon present)	97.5	11	$104,144	91

Source: Cohen, 1988:553.

table 7.2 are only preliminary and are intended only to stimulate further research and analysis.

Cohen's final example of possible applications of monetary estimates of harm relates to the problem of constructing an empirically based sentencing guideline system. He notes that U.S. Sentencing Commission's federal guidelines (1987) are based on the current average sentence for various types of offense-offender characteristics. Thus the median sentence for a "basic" bank robbery when no weapon is present is thirty-seven months. The sentence is then adjusted upward as other factors are added to the offense, such as the possession of a weapon or bodily injury. These sentences are based on current practice.

Cohen suggests an alternative approach involving estimation of the risk of injury and death associated with each stage of a bank robbery. These risks could then be converted into monetary equivalents based on cost estimates. This would provide an alternative estimate of the severity of each stage of the robbery that could then be compared to current practice. Table 7.3 is designed to illustrate how the monetary estimates of harm could be helpful in constructing sentencing guidelines.

The first column shows the median sentence length under the federal guidelines. The second column shows the marginal increase in sentence length. The last two columns display similar information for monetary estimates of harm. Cohen does not argue that such estimates should be used directly to determine sentence lengths but rather that they provide information to be taken into account. In the case of bank robbery, he says that the estimates suggest the need for an increase in the marginal penalty for carrying a weapon in a bank robbery as well as for the more serious bodily injury categories.

Critique of Cohen's Study

As these applications illustrate, Cohen offers his victim cost figures as a means of comprehensive arithmetic for criminal justice decisions that range from cost-benefit for crime control to a rather utilitarian measure of just punishment. With such broad ambitions, however, the victim cost schema presented is vulnerable to a number of objections. The specific cost estimates are opportunistic, arbitrary, inconsistent, and too high. The schema lacks an articulated theory of either public or private costs of crime and provides no indication of the connection between public and private cost. Moreover, Cohen's analysis reveals no relationship between its cost estimates and its conclusions about the cost-effectiveness of the investment in further crime control resources. These three deficiencies in the analysis are in fact typical of those found in other similar exercises that have appeared in the literature rather than being idiosyncratic. As such they merit close examination.

Specific Cost Estimates

The Cohen estimates are opportunistic because Cohen appears to have searched for various existing monetary measures to fit some attributes or consequences of crime and, whenever he could find them, plugged them into his crime cost equations. He provides no list of criteria by which particular measures such as pain and suffering damages in tort cases or wage differential estimates were either selected or judged appropriate.

When no monetary equivalents exist for certain consequences of crime, they are ignored. Obviously criminal behaviors that are more closely analogous to previously measured and monetized attributes or consequences will by this method emerge as more expensive than other offenses. Thus neurosis stemming from rape becomes a multiple-thousand-dollar cost, whereas the fear of homebreaking associated with burglary is apparently downplayed because of the lack of physical contact. The observer is left to speculate about the costs of treason, criminal conspiracy, and contributing to the delinquency of a minor, because of the lack of monetary cost analogies for those offenses. The same is true for large offense categories such as illicit drugs.

If Cohen's methodology is opportunistic, his choice of particular measures for particular factors seems wholly arbitrary. Jury awards are the basis for pain and suffering estimates because he selects a concept of cost from the tort system rather than workers compensation schemes. Why is this? Having chosen a tort measure for pain and suffering, he rejects a tort measure for death costs. Why is a wage differential measure of death costs superior to a wrongful death tort measure? Who in Cohen's analysis suffers the economic costs associated with the risk of death from crime: crime victims or their families?

Without any statistical criterion for choosing among different cost measures, the number of conclusions we can come to about Professor Cohen's cost choices are limited. We can say that there are manifest inconsistencies in moving across systems and that the selection in each of his major categories appears quite arbitrary. We can also guess that the selections were made to generate very high cost figures for most categories of crime.

Pain and suffering damages for personal injury in Anglo-American law are notorious for both their arbitrariness and their inflated size. In fact these inflated findings are the principal target of no-fault insurance reform and a variety of other efforts to displace them in accident compensation plans. Yet Cohen wants to adopt these measures as appropriate measures of crime costs.

It is interesting that no scholar has ever attempted to determine the cost of automobile accidents by multiplying the number of accidents by the jury award for the average pain and suffering experienced, though that is precisely what jury awards are supposed to represent (see Conard, 1964). Instead these jury award averages, profits without honor in their own legal categories, are being used to estimate the pain and suffering costs of crime. The use of these figures certainly makes crime appear more costly, but the use of a similar multiplication would also make automobile accidents, industrial accidents, and diseases appear vastly more costly. The rank order of costs associated with occurrences will only change between painful and nonpainful phenomena. For all phenomena involving bodily injury, using inflated pain and suffering estimates simply adds zeros to the aggregate totals.

A pointed example of this inflation without policy significance concerns the selection of an estimate of $2 million per life lost for criminal homicide that Cohen tells us was derived from wage rate differential studies (1988:548). Of course, certain persons would pay such a price or much more to avoid lethal consequences. However, the aggregate costs produced by such estimates are unrealistically in excess of the resources that any society would make available to prevent crimes carrying such consequences. If the average cost of a death is $2 million, the 2.16 million deaths experienced in the United States in 1990 generate an aggregate "cost" of $4.32 trillion (2.16 million × $2 million), approximately 5 percent more than the $4.12 trillion gross national product of the United States in 1989 (see figure 7.1).

It might be objected that selecting a $2 million cost figure from a wage rate differential study and applying it as an across-the-board estimate per death in regard to more than 2 million deaths from all causes in the United States is a nonsensical exercise. The more than 2 million persons who died in the United States in 1990 were all of different ages, health conditions, and levels of economic activity. Generalization of the cost of each of their deaths at $2 million per occurrence without any attention to the specific circumstances of the victims seems totally

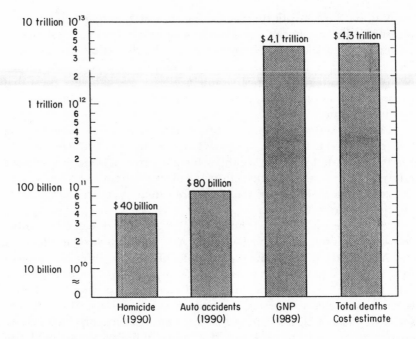

Figure 7.1. Hypothesized Mortality Costs and Total Economic Activity in the United States, 1989–90.

unjustified. This kind of criticism is unavailable to proponents of Cohen's analysis because that sort of undifferentiated aggregation without any reference to the circumstances of the victims is precisely his methodology.

If one were to construct a list of the harmful consequences experienced by all the members of a society, the aggregate social dollar-cost measures of unpleasant consequences would always far outstrip the aggregate value of the economic output of that society. What would be the total national cost of pain and suffering generated by disease and injury? If the aggregate costs of social harm are vastly in excess of available economic resources, the money cost estimate for any individual harm cannot justify a claim for that level of resources as a public investment in prevention or remediation. At best the relative costs of different sorts of harm could provide guidance about the appropriate relative scale of investment of public resources in the prevention of different sorts of harm. Perhaps, then, if cancer generated four times the cost of some other disease, this might argue for larger public investment in the more costly problem.

In summary, the particular cost estimates of crime in Cohen's study overemphasize what has been quantified; are arbitrary in their choice of cost standards in particular cost categories; are inconsistent in their choice of cost criteria when moving from categories such as pain and suffering to loss of life; and wildly overstate the level of resources that could or should be directed at crime in general or at any particular crime.

The Theory Gap

Later in this chapter we will demonstrate that there are many different ways to conceptualize the social costs or public costs of crime. What is remarkable about Cohen's analysis—and that of Zedlewski, which we will examine next—is that no explicit theory for the social cost of crime is expounded or justified. Yet to try to estimate aggregate cost totals before defining what one means by the cost of crime is to put the cart before the horse.

Let us give two hypothetical examples of the problems generated by uncertainty about the definition of social cost. Smith owns a Mercedes Benz that recently cost her $50,000. She parks the car on the street rather than pay $75 per month for locked garage rental as an antitheft measure. Nor does she use an automobile theft deterrent system that she could purchase for $200, nor is her car insured against theft. When the vehicle is stolen Smith's personal loss is $50,000, but what is the social cost of the crime? Parts of Cohen's analysis seem to suggest that the cost of the car theft is $50,000 and that any public expenditure to prevent it up to $49,999 would be justified on a cost-benefit basis. Without an explicit definition and discussion of social cost, we can only guess at this conclusion from the way in which he treats other problems.

Jones, by contrast, does not lose his car by theft but, after consuming a large quantity of alcohol, drives it into a concrete lamp post and dies in the collision. What is the social cost of this crime? Does it include the $2 million estimated value of Jones's life? If so, all of the 20,000 or so annual alcohol-related fatalities should be added to the cost of crime, and up to $40 billion should be devoted to preventing drivers from causing the deaths of themselves as well as others. Moreover, that alcohol-related fatalities are more likely to be self-inflicted than other fatal accidents should be no reason to reduce the relative level of preventive resources devoted to them. We cannot tell how Cohen might wish to address these questions.

These are not small matters. Most traffic accidents involve infractions of the law and these alone could triple the cost of crime as Cohen proposes to measure it. Two-thirds of the crime problem in the United States as measured by Cohen should in fact be traffic accidents, a phenomenon that should lead to rather different criminal justice and punishment priorities than currently obtain.

Beyond the practical significance of different definitions of social cost, Cohen's analysis never addresses certain basic conceptual problems because it does not confront the simple but significant issue of when and how society as a whole should regard the harm and losses suffered by the victims of crime as collective rather than as individual problems.

The Fallacious Use of Cost-Benefit

A return to the case of Ms. Smith and her unguarded Mercedes Benz reveals a further flaw in Cohen's analysis. He mistakenly assumes that any course of conduct that appears to be cost justified when the cost of crime is compared to the cost of using prisons to prevent it, is therefore also cost effective in a real-world setting that may offer other policy choices. Spending $49,000 for prison rather than losing a $50,000 car might seem like a good idea if there were only two choices, but in a world with many cheaper methods of prevention available—the locked garage and the automobile theft deterrent mentioned above—it represents an enormous waste of resources. It is true that there is one sentence referring in the final section of Cohen's paper to the possibility that "alternatives to incarceration could yield higher benefit-cost ratios" (Cohen, 1988:551), but this seems restricted to alternative punishments and the matter is not pursued. The theoretical and practical issues raised by alternative methods of crime prevention are ignored by Cohen in a way that invalidates his whole discussion of cost benefit.

Prison Costs versus Release Costs (Zedlewski 1987)

Edwin W. Zedlewski's brief monograph *Making Confinement Decisions* appeared in 1987, the year before Cohen's study. He had previously published (*Public Administration Review*, 1985) an article entitled "When Have We Punished Enough?" in which he had argued, from "a cost-benefit perspective," that the available data suggested "that greater social benefits are derived from prison incarceration than are usually assumed." The results of his analysis, he added, "overwhelmingly support the case for more prison capacity" (Zedlewski, 1985:771, 778). He acknowledged, however, that existing data were adequate "only for a crude answer" to the title question.

In *Making Confinement Decisions* Zedlewski refined and expanded his argument and presented a more detailed critique of what he called "our reluctance to incarcerate" (1987:2). The conclusions of Cohen's and Zedlewski's analyses are not discordant but are arrived at by different routes. The emphasis in Cohen's study is on victim costs. Zedlewski is concerned with "social costs based on money spent." Although Zedlewski claims that "increasing prison capacity is likely to save communities' money by averting a variety of costs imposed by crime," he says that "what actual savings would be realized is open to speculation." Phenomena such as "victim harm" and "public fear" are said to "defy quantification." "Some savings of victim losses would surely result [from "increasing prison capacity"]" but such things as "costs incurred by victims of violence are difficult to express in dollars" (Zedlewski, 1987:2, 5).

One of the major differences in approach between Cohen and

Zedlewski is that in the case of the latter, the emphasis throughout is on "the appealing concept of incapacitation" and "the case for more prison capacity" (Zedlewski, 1987:5, 6). Cohen also writes about "incapacitation policy" but is cautious about the implication of his "crime specific estimates." Thus he says that although the estimates in table 7.2 suggest that longer prison sentences are warranted for rape, assault, and auto theft, "it must be emphasized that these are *not* policy recommendations" and "furthermore, it is possible that alternatives to incarceration could yield higher benefit-cost ratios" (Cohen 1988:550–51).

Zedlewski, by contrast, does a benefit-cost analysis without employing any crime-specific estimates and makes recommendations regarding the average length of sentence for all crimes in general. Moreover, he is quite unequivocal about the policy implications of his analysis. One year's imprisonment is said to involve costs of $25,000, whereas the social costs averted by that imprisonment through incapacitation are estimated to be $430,000, or more than seventeen times as large (Zedlewski, 1987:3, 4). Whereas the emphasis in Cohen's analysis is on "the cost of crime to victims" (Cohen, 1988:545–49), Zedlewski's analysis focuses on what he refers to as "release costs."

Zedlewski is primarily concerned with "the release of borderline offenders—those offenders who would have gone to prison had space been available" and in particular "the social cost incurred by releasing these offenders" (1987:3). Release costs are approximated by estimating the number of crimes an offender is likely to commit if released and multiplying that number by an estimate of the average social cost of crime. One of the most significant features of Zedlewski's analysis is that the offender is virtually a cipher. His "typical inmate" and "borderline offender" are virtually indistinguishable. "Many releasees," he says, "are likely to be more criminal than some who are imprisoned" (1987:3).

Cohen distinguishes between offenses and suggests that "longer prison sentences are warranted for rape, assault, and auto theft, and that shorter sentences might be warranted for burglary and larceny" (1988:551). Zedlewski's analysis is in terms of such concepts as the "convicted offender," the "typical inmate," and the "typical prison-bound criminal" who is responsible for an "average number of crimes committed in a year." He is not concerned with distinctions between types of offense or offender.

Thus "we find that a typical inmate . . . (committing 187 crimes per year) is responsible for $430,000 in crime costs." It follows therefore that "sentencing 1,000 more offenders (similar to current inmates) to prison" could only be beneficial. In fact "about 187,000 felonies would be averted through incapacitation of these offenders." What Zedlewski is recommending is a penal strategy of *collective or general incapacitation*. By contrast, an example of the policy application of victim cost estimates given by Cohen relates to an "incapacitation policy" directed specifically at robbery offenders—that is, a somewhat more *selective incapacitation* policy—which he estimates "would have a benefit-cost ratio of about 3 to 1" (1988:550).

Critique of Zedlewski's Study

We have previously commented on the claims made in Zedlewski's paper (Zimring and Hawkins, 1988; Zimring, 1989), and we will not repeat here all of the issues dealt with in that prior review. However, it is worthwhile to compare the approaches of Zedlewski and Cohen and to examine the extent to which the claims made regarding dollar costs and dollar benefits in a study like Zedlewski's relate to arguments that could be (better) based on the same data without dollar cost transformations.

When compared with Cohen's analysis, the Zedlewski argument is based on different but equally opportunistic and arbitrary measures of crime costs. Zedlewski excludes the two major money cost categories in the Cohen analysis: death risk and pain and suffering. So that the aggregate cost of crime in his analysis is, in 1983 dollars, about one-quarter of that claimed by Cohen. These omissions not only influence the aggregate cost figures for crime and the cost-benefit ratios that can be claimed for imprisonment, they also have a profound influence on the rank ordering of costs for different types of crime.

For Cohen even the most painless homicide generates a cost—$2 million—equal to more than 10,000 times the cost of a personal or household larceny, and more than 1,500 times the total cost of a burglary. For Zedlewski all crimes are created equal at the $2,300 cost figure. Although part of the difference is the result of the single-category aggregation in his argument, much of the difference also results from Zedlewski's analysis leaving the principal features of crime loss in Cohen's schema uncounted.

Indeed Zedlewski could be regarded as a somewhat miserly for estimating only $430,000 in savings for every additional person-year of imprisonment in the 1980s. Assuming that his "187 crimes" was a valid estimate of offenses avoided, the use of cost categories that followed Cohen could result in savings from a year's incarceration in excess of $1.5 million, and an extra 1,000 prisoners incarcerated should produce savings in the range of $1.5 billion. In a state like California, where the total prison and jail population rose by 100,000 in the 1980s, this increase should have produced savings in social expenditure on crime during that decade of $150 billion, or three times the government budget for all causes.

But since Zedlewski's incapacitation estimate would predict that the first 12,000 to 20,000 additional prisoners in California during the 1980s would have driven California's crime rate down to zero, it is difficult to understand where the last $120 billion in crime-saving benefits would come from. Figure 7.2 shows the impact of extra prison and jail popula-

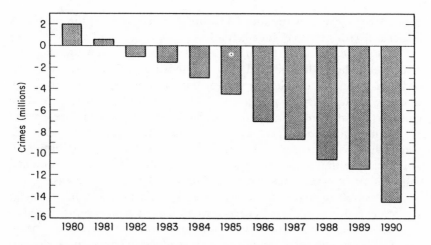

Figure 7.2. Hypothesized Crime Volume in California, Using 1980 Crimes, the Zedlewski Model of Incapacitation, and California Prison and Jail Trends.

tion crime rates in California at 187 crimes per extra prisoner year: by 1990, the extra jail and prison population in California would have eliminated crime ten times over if the Zedlewski estimates had held.

In one respect, the Zedlewski analysis of crime costs is superior to that of Cohen because it embodies some explicit theory regarding the social cost of crime. Zedlewski estimates the social cost of crime by adding to the direct victim losses due to crime all the expenditure for activities that have as a major element in their justification a response to, or the prevention of, crime. There is here again no rationale for counting all private costs as social costs.

The implicit assumption that a zero crime rate would put an end to expenditure in the United States on police and large dogs is also open to question. Another debatable assumption is that fluctuations in the crime rate automatically result in a proportionate response in crime prevention expenditures so that if the aggregate crime rate decreased by 25 percent, police budgets would be proportionately reduced. There is no evidence of anything of that nature ever occurring anywhere in the world.

A principal problem with Zedlewski's estimate of imprisonment benefits is the inflated estimate of 187 crimes per year saved with the incarceration of each additional prisoner at the margin between release and confinement in existing criminal justice policy. The Rand Corporation researchers, whose report is the only evidence Zedlewski cites, would regard 187 crimes as an overestimate by a range of between 1,000 and 2,000 percent, for reasons that were explored in chapter 5. Of significance here are the gains and losses that occur when both crimes and criminal justice measures are translated into dollar terms in studies like these. We believe that the monetization of crime adds to the problems associated with studies of incapacitation.

Not among the benefits of the use of dollar equivalents to the costs of crime in the Zedlewski analysis is any sense of which offenses should be regarded as more or less serious by the community, for all that is produced is a single aggregate figure of $2,300. While this figure must be close to the value that someone using Zedlewski's logic would determine for the high-volume property crimes that dominate aggregate crime statistics, there is no reason to believe that the $2,300 value represents an appropriate estimate for any single crime and there is no way to obtain a value for any specific crime from the data set used in the analysis.

The single apparent benefit of the monetization of crime costs in this study is that it provides a basis for comparing what is represented as the social costs of imprisonment in the same dollar terms as social benefits. The advantage must lie in the increased significance to be derived from saying that "a year in prison [at] total social costs of $25,000" would produce a saving of "$430,000 in [reduced] crime costs." The question is whether that provides any additional information over and above concluding that "the social cost of an imprisonment decision—about $25,000 per person year"—would save the community from 187 crimes that would otherwise be committed (Zedlewski, 1987:3, 4). Given the arbitrary and unjustified assumptions involved in deriving the dollar cost estimates, it may be that the latter statement conveys more meaning to the reader because it is much easier to imagine what the prevention of 187 miscellaneous offenses might represent than it is to comprehend the $430,000 estimate that Zedlewski assembles.

However, the monetization of crime costs and benefits does have tactical advantages in Zedlewski's analysis that are significant disadvantages to cost considerations in crime control. Converting crimes into dollars seems to relieve Zedlewski of the burden of defending the evidentiary base for his 187 crime prevention savings. Instead, after the conversion Zedlewski never returns to the question of crime volume but simply continues to speak in terms of dollar volume when defending his argument about prison costs.

> By combining crime costs and offense rates, we find that a typical inmate in the survey (committing 187 crimes per year) is responsible for $430,000 in crime costs. Sentencing 1,000 more offenders (similar to current inmates) to prison would obligate correctional systems to *an additional $25 million per year.* About 187,000 felonies would be averted through incapacitation of these offenders. *These crimes represent about $430 million in social costs.* The conclusion holds even if there are large errors in the estimates: Doubling the annual cost of confinement, halving the average crime per offender, and halving the average cost per crime would indicate that $50 million in confinement investments would avert $107 million in social costs. (Zedlewski, 1987:4; emphasis added)

The second specious advantage of monetization is that the dollar figures seem to assume a near empirical reality for their author so that they end up substituting for, rather than supplementing, the hard evidence on the impact of crime control measures on crime rates. The lengthy quotation in the previous paragraph is an arithmetic excursion with no basis in fact, in Zedlewski's monograph, or in any of the references he cites. We have here a factitious two-step process in which crimes are converted into dollars, dollars are manipulated on the basis of questionable assumptions into arithmetic formulas to observe how they affect the bottom line, and the entire process is presented as empirical research.

A significant drawback to the monetization of crime in studies like Zedlewski's is that conversion into dollars in effect changes the subject from crime prevented to hypothesized dollar savings. Once hypothetical dollars have replaced crimes as a unit of analysis, the manipulation of dollar values in arithmetic formulas becomes the principal focus of analysis. We cannot say that monetization is the cause of the sort of ludicrous overreach evident in the Zedlewski paper, as Zedlewski may be capable of repeating the claim of 187 crimes per prisoner year without the protective aid of dollar value measurement. We do know that extreme claims in the literature are associated with monetized analytic strategies (see, e.g., Abell, 1989; Gramm, 1993).

Furthermore, insupportable incapacitation estimates seem to have a longer shelf life and a greater invulnerability to criticism once they are converted into dollar terms. Whether the conversion into dollar cost figures is a major cause of error and overstatement or merely contributes to a general environment in which misstatement thrives, the absence of any significant benefits associated with dollar cost conversion makes a cost-benefit analysis of its use an easy matter.

Yet if previous efforts have been associated with overstatement of epic proportions of the monetary costs of crime, it remains to be seen whether the problems involved in dollar cost estimates are inherent or avoidable. To pursue this issue it is necessary to look beyond current attempts at cost analysis and consider the issues of theory that arise in relation to even the most straightforward attempts to convert crime into dollar cost figures.

III. Problems in Social Cost

Though ignored in current writings like that of Cohen and Zedlewski, the subject of the cost of crime has received attention in twentieth-century criminal justice policy studies. For many decades interest in the topic followed a cyclical pattern. Major investigations of criminal justice would focus attention on the cost of crime but then would encounter theoretical difficulties that would lead to the abandonment of ambitions to make comprehensive assessments of cost. There matters would rest until the next major research project renewed the cycle. This pattern is exemplified in the reports of the National Commission on Law

Observance and Enforcement (the Wickersham Commission) published in 1931; the Cambridge University's Institute of Criminology's
aborted research project in the 1960s; and the U.S. President's Commission on Law Enforcement and the Administration of Justice in 1967.

The earliest attempt to deal with this matter systematically was
made by the Wickersham Commission over sixty years ago. By the time
the commission completed its investigation and published its 657-page
report on *The Cost of Crime* (1931) under the authorship of G. H. Dorr
and S. P. Simpson, the prognosis for measuring the cost of crime looked
grim. The commission concluded that

> [i]t is wholly impossible to make an accurate estimate of the total eco-
> nomic cost of crime to the United States. This is true whether we look
> at the immediate cost of crime to the tax-paying and property-owning
> public and the individuals composing it, or whether we consider the
> net ultimate cost to the community as a whole. Many "estimates" of
> total cost have been made, but they, in our opinion, have only been
> guesses; and we do not feel that any useful purpose would be served by
> still another guess. (National Commission, 1931:442)

Subsequent commentary on the question was no less pessimistic. In
1936 Hawkins and Waller published their "Critical Notes on the Cost of
Crime" quoted at the beginning of this chapter. Although they com-
mended the Wickersham Commission report as "the most completely
thought out discussion of the cost of crime," they concluded their critical notes by stating categorically, "we cannot measure the cost of crime"
(Hawkins and Waller, 1936:679, 694). In 1942 Frederick Conrad, in
"Statistics in the Analysis of Social Problems," noted that published "estimates on the annual cost of crime vary from 1 to 18 billion dollars".
After pointing out a number of examples of "discrepancies, inaccuracies, duplications and assumptions in the figures," Conrad concluded
that "in view of the difficulties involved in attempting to measure social
problems by monetary standards the elimination of cost data should be
considered and attention given to other methods of approach"
(1942:540, 549).

More than three decades of important work in economic theory
separated the work of the Wickersham Commission from that of the
University of Cambridge's Institute of Criminology, which in 1964 published "Design of a Study of the Cost of Crime" under the authorship J.
P. Martin and J. Bradley. The authors paid tribute to the "considerable
virtues" of the Wickersham Commission's "major study of the cost of
crime" but also noted some of its limitations. They announced, however, that it was now possible "thanks to the development of social
accounting and survey techniques . . . both to make a more refined
analysis of some topics covered by the Wickersham Commission and to
tackle questions beyond the power of the methods available in 1930"
(Martin and Bradley, 1964:591–92).

It was therefore proposed that as part of its research program the institute was going to conduct a survey into the cost of crime. This would cover public expenditure on law enforcement, the administration of justice, the treatment of offenders, and also the various costs that were privately borne. The aim of the study would be to make a systematic analysis of expenditure in the relevant public services, and from private sources in terms of both the current pattern and of the trends leading up to it. The researchers planned to start with an analysis of public expenditure and then deal with the private sector as resources of staff and money became available.

Subsequently, however, in a paper entitled "The Cost of Crime: Some Research Problems" by J. P. Martin, a less sanguine note was sounded. Martin reported that "an apparently simple subject is in fact one of great complexity." He set out at length some of the general methodological and conceptual problems involved in computing the cost of crime. He noted that

> the concept of public cost . . . depends on the making of various assumptions and, if these become too numerous or too cosmic, the value of the resulting figures may be minimal; [it] tends to involve guesswork on such a grandiose scale that the result might as effectively be obtained from the study of a crystal ball. (1965:57–58)

Although it was suggested that the subject might be "divided into a series of relatively self-contained sections each of which could be the subject of a separate project or sub-project" (Martin, 1965:57–58, 63), that appears to have been the end of the Cambridge Institute's investigation of the cost of crime.

Our final example is taken from the report of the U.S. President's Commission on Law Enforcement and Administration of Justice published in 1967. "Crime," the commission reported, "affects the lives of all Americans [in] that it costs all Americans money . . . crime in the United States today imposes a very heavy economic burden upon both the community as a whole and individual members of it." At the same time the commission noted that the available cost information was fragmentary. "The lack of knowledge about which the Wickersham Commission complained thirty years ago is almost as great today."

Though some cost data were reported in the Uniform Crime Reports and some were available from individual police forces, insurance companies, industrial security firms, trade associations and others, the commission found that "the total amount of information is not nearly enough in quantity, quality or detail to give an accurate overall picture." Under the circumstances the commission concluded that "the information available is most usefully presented not as an overall figure, but as a series of separate private and public costs."

Accordingly, the commission presented some figures relating to what it referred to as "six different categories of economic impacts both private and public." Yet even in respect to this "picture of crime as seen through cost information" the commission acknowledged serious deficiencies:

> Numerous crimes were omitted because of the lack of figures. Estimates of doubtful reliability were used in other cases so that a fuller picture might be presented. Estimates do not include any amounts for pain and suffering. Except for alcohol, which is based on the amount of tax revenue lost, estimates for illegal goods and services are based on the gross amount of income to the seller. (Gambling includes only the percentage retained by organized crime, not the total amount gambled.) The totals should be taken to indicate rough orders of magnitude rather than precise details. (1967:32)

The only recommendation the commission made about the cost of crime was "that the lack of information about the economic costs of crime in America be remedied." However, despite this almost ritual incantation in the field of criminal justice about "the need for more data," it was clear that once again the pursuit of "an accurate overall picture" of the cost of crime had been abandoned (President's Commission, 1967:31–35).

Recent treatments of cost and benefit have diverged from the cyclical pattern we have noted. Discussions of the cost of crime and of cost-benefit analysis show no signs of abatement. Does this mean that the theoretical problems encountered by previous investigations have been finally resolved? Or are they merely being ignored in contemporary literature on the cost of crime? We tend to support the latter view.

Two Problems

Two relatively simple illustrations can lead us to many of the basic definitional problems regarding the social cost of crime. First is the hypothetical example set forth in section II in which a $50,000 car is stolen and the owner is not insured against its loss. The second illustration concerns illegal gambling transactions in which a customer of an illegal sports bookmaker places $10,000 in bets and $7,000 of the $10,000 is a net loss to the customer.

Theft

In the first case, that the monetary loss to the owner of the stolen Mercedes Benz was estimated at $50,000 poses no problem. One does arise, however, when we try to translate the individual's loss into the net social cost of the theft. Both Zedlewski and Cohen assume that the individual loss and the social cost are the same amount. The standard economists' objection to this assumption has been clearly stated by Philip J. Cook in reference to the illegal taking of money or goods from an unwilling victim.

Viewed dispassionately, . . . in this type of crime the victim's monetary loss is equal to the criminal's monetary gain; on balance then, there is no direct social cost to the transfer itself . . . From the larger perspective of society, the direct social cost of such crimes is equal to the loss of the criminal's time (as well as the other resources he devoted to perpetrating the crime), which would otherwise have been used in legitimate activity. (Cook, 1983:374; see also Becker, 1968:170–71)

Since some member of society is obtaining use of the car, the social-cost calculation will be considerably lower than $50,000. One objection to this treatment of cost is that the transfer of benefit from the victim to the thief is a palpable social loss that should be recognized. One way to do this is to define the theft as outside the boundaries of the social network in which the use of goods and services is counted as gain. As Robert Gillespie puts it, "society could just as well be defined as being composed of only noncriminals, in which case transfers would clearly impose a social cost" (Gillespie, 1976:6).

Two points can be made about the choice of cost logic. First the consequence of choosing between the narrow view of Cook following Gary Becker (Becker, 1968) and the broader Cohen-Gillespie view of the car theft can be substantial. Car thieves do not often have high-paying alternative employment available, so the social cost figure will vary between $50,000 and a few hundred dollars. On a tight reading of social costs like Cook's, an act of vandalism may be more destructive than the theft of more valuable property. Second, there is no single concept of "social cost" that will automatically lead to a correct answer here. There are instead different ways of emphasizing the consequences of theft, including one that stresses the loss to the community and another that emphasizes the loss to individual victims. With such differences outstanding, using the cost of crime to determine criminal punishments or law enforcement resources seems a vain hope.

Gambling

Neither is the social cost of gambling easy to determine. In the example we just gave one might argue that the individual cost of the criminal act to the client placing the bet was the $7,000 lost. His capacity to purchase other goods was certainly diminished by that amount. The problem with this case is that the customer got what he paid for. As Cook has pointed out in relation to all "consensual crimes": "Since their participation in such activities is voluntary the perpetrators evidently believe that the benefits derived from such activities outweigh the cost" (1983:376). On that analysis no individual loss can be derived from any measure of expectation or loss.

What about social cost? The Report of the President's Commission on Law Enforcement and the Administration of Justice in 1967 estimated the cost of illegal gambling in 1965 dollars at $7 billion, using a formula in which the losses of individual gamblers—net of their gains—was the selected measure of "economic cost." Though the commission did not justify its choice of measure, this sort of "balance of payments" criterion is consistent with regarding organized crime as outside the pale of the legitimate economy and therefore its net income from gambling as a social cost. This approach poses problems for a number of reasons, not least because the criminal gambler's purchase of other goods in the legitimate economy is not counted as an offset in this accounting of social cost. Perhaps the justification that would be offered for this is that the transfer of economic power to social enemies in some sense represents a social cost.

The implications of this conclusion about gambling on the President's Commission's calculations were astonishing. The economic cost of gambling was estimated at nine times the economic cost of homicide in 1965 and far exceeded the commission's estimated cost of all reported and unreported theft, fraud, and embezzlement in the United States. When added to other revenue from consensual crime the commission reported that "Organized crime takes about twice as much income from gambling and other illegal goods and services as criminals derive from all other kinds of criminal activity combined" (President's Commission, 1967:32). However, Cook disagrees:

> It should be noted that the measure of the direct social costs of illegal gambling used by the President's Commission on Law Enforcement and Administration of Justice—revenues net of payoffs—is not appropriate as a measure of social cost, although it does give an indication of this industry's contribution to the gross national product. (1983:376)

The same could be said for the other categories of vice income, such as drugs and commercial sex.

Neither the President's Commission nor Cook were in any hurry to draw policy conclusions from this kind of cost analysis. Though the President's Commission decided that gambling generated costs nine times those of homicide, its only recommendation was "that the lack of information about the economic costs of crime in America be remedied" (President's Commission, 1967:35). Cook remarks in this connection that "Presumably legislators outlaw these consensual activities because their constituents view them as immoral and destructive; that is people who are not directly affected by the consumption of these commodities object to them on moral and paternalistic grounds" (1983:376). Yet the political consequences of regarding consensual crimes, in which all parties directly involved are voluntary participants in the illegal activity, as costless would logically be zero investment in law enforcement, not only in respect to gambling but also in relation to a wide range of vice-related goods and services such as illicit drugs and prostitution.

The confusion and uncertainty regarding the status of consensual crimes have enormous implications for any attempt to construct an aggregate cost schedule involving the costs of criminal activity. Either consensual offenses are twice as important as all other types of offenses or they should be totally disregarded. Yet an aggregate figure that is the product of a computation of the cost of crime for all categories based on either of those premises would be problematic in the extreme. And the issue is not confined to consensual crimes. Measuring the total social harm generated by that behavior as the efficiency costs of criminal behavior, is nonsensical, for reasons that are easy to state.

The Economist's Dilemma

The problem with a definition of the cost of crime solely concerned with efficiency is that it is so frequently irrelevant to criminal justice policy. Inefficiency is not a crime. There are many social practices that impair efficiency which are not regarded as fit subjects for criminal prohibition, and there are many economically efficient practices that are considered harmful enough to demand criminal classification.

It is true, as Hawkins and Waller noted in 1936, that "the distinction between crime and 'productive' labor is a moral and not an economic distinction" (1936:684), but many wish to prohibit the activities of the drug dealer, the pimp, and the bootlegger regardless of their impact on national income. The criminal law is a moral system, *not* an economic one.

Thus, lurking just behind the economist's response to the problem of social cost in consensual crimes is a dilemma that confronts economic concepts of the social cost of crime. Either economists restrict their notion of social cost to efficiency costs—in which case the concept cannot measure many of the social harms that inspire criminal prohibition—or analysts try to expand the conception of cost to cover other elements of harm, in which case the concept loses any precise economic meaning or coherence. Thus economists become irrelevant to criminal justice policy if they adhere to efficiency costs and incoherent if they do not.

This does not mean that the economic analyst can play no useful role in criminal justice policy, but it does mean that economic concepts of cost cannot be comprehensive guides to punishment policy without a revolutionary change in the foundations of criminal law. This is a strategic problem far broader than the specific missteps of Cohen and Zedlewski. One cannot do a cost-benefit analysis of laws not based on economic criteria. The wonder of the current situation is that monetized values for the "cost" of crime can obscure this simple point.

Conclusion

Criminal justice studies took a significant step backward in the 1980s, when arguments about public costs and the criminal incapacitation benefits of imprisonment were transformed into propositions that attempted to express both the costs and benefits of imprisonment in dollar terms. The major strategic problem with this approach is that many of the most significant harms associated with crime have nothing to do with economic efficiency. Moreover, others can neither be well measured nor easily translated into public costs. A further problem is that monetized cost estimates add nothing to the proper calculus of choice between the costs of imprisonment and its benefits. Once we know the kind and amount of crime prevented by particular penal measures, the choices to be made can be as rationally discussed in the terms in which they naturally occur.

The only defensible cost-benefit analysis in dollar terms would be confined to the economic impact of crime. This might provide helpful partial guidance for some criminal justice decisions, but it is a singularly inadequate way of addressing the harms associated with most types of crime. Dollar cost figures for crime can thus either represent a complete non sequitur or a very limited guide to policy decisions.

Certainly occasions will arise when relative resource costs can be a useful guide to policy, as when a choice has to be made between two equally effective crime control measures. There may be other occasions when the differential economic impact of two otherwise similar criminal offenses can guide us to the appropriate candidate for priority in law enforcement.

Well over half a century ago, Hawkins and Waller concluded their "Critical Notes on the Cost of Crime" as follows:

> The significant point to be emphasized by the present analysis is that instead of attempting to discover the cost of crime, an enterprise foredoomed to some absurdity, we need to know the nature and magnitude of the probable immediate results of a crime crusade. We need to be more cognizant of the permanent consequences of crime as an organic part of our society . . . What unintended consequences for the larger social order have such crimes as bank robbery, embezzlement, counterfeiting, and racketeering? Who pays for goods that have been stolen or destroyed; what problems of shifting and incidence appear? What are the roots of crime in legitimate business? These, it seems to the present writers are some of the significant questions that arise from a study of the cost of crime. It is within the power of the human mind to answer them. (1936:694)

Our own more recent adventures in cost and benefit only confirm the soundness of their conclusion and remind us once again of the tendency for history to repeat itself.

8

Incapacitation and Imprisonment Policy

The aim of this chapter is to address directly the relationship between incapacitation as a penal purpose and imprisonment policy. We intend this as a valedictory analysis, drawing on the issues developed in the previous chapters to consider the linkage of purposes with policy choices in contemporary criminal justice. Section I attempts to put incapacitation as a motive in the broader policy contexts of crime control, criminal punishment generally, and imprisonment. Section II contrasts the different roles of incapacitation in two offense categories—drugs and property crime—where the choice between prison and other sanctions is of particular importance in imprisonment policy. Section III examines how different processes that may impose limits on the size of prisons affect the way in which incapacitation influences the size and character of a prison population. A concluding note predicts two future developments in the impact of incapacitation on penal policy.

I. Incapacitation in Its Contexts

Our objective in this section is to place incapacitation in the context of the broader policy environment that surrounds a range of questions about crime and criminal justice policy. How does incapacitation relate to the great variety of different strategies available to control crime? What is its relationship to the other commonly listed purposes of criminal punishment? What is the properly understood relationship between the practice of imprisonment and the objective of incapacitation? The path that this section takes is from the broader environment of crime control in general to the more restricted territory of the purposes of punishment and on to the still more specific domain of the modern prison. The central distinguishing characteristic of restraining sanctions is physical control. The most significant policy choice in imposing criminal sanctions is between monitoring the behavior of subjects at risk in community settings and responding to community risk by isolation and physical control.

Restraint and Crime Controls

Restraint through penal confinement is only one among a large number of different strategies for preventing crime. Comparing the incapacitation strategy with other categories of crime prevention is important both to place that strategy in a larger context of policy options and to demonstrate those aspects of incapacitation that give it a preferred position as a method of prevention for serious crime.

The range of crime prevention strategies that can be used to prevent or provide defense against crime includes nonrestraining punishments intended to deter and stigmatize; environmental controls to reduce opportunities for crime; and a number of methods of monitoring individual behavior in ways that reduce personal opportunities to commit crime and make the identification of responsible offenders easier. Each of these major categories contains a large number of crime prevention mechanisms. Nonrestraining punishments include fines, loss of privileges and status, community service, and so forth. Environmental adjustments vary from cashless buses to steering-wheel ignition locks for automobiles, security codes and identity checks for financial transactions, locks, big dogs, and deliberate behavioral change on the part of persons at high risk of victimization. Nonrestraining mechanisms to control and monitor the behavior of potential offenders range from bank video cameras and periodic drug testing to probation and parole supervision.

Two factors account for most of incapacitation's preferred position as a crime control strategy. First, restraining techniques operate upon the convicted offender who is presumed to be at risk of committing future crime. Second, restraint directly controls the behavior of the potential offender rather than leaving him or her any choice in the matter. Restraining mechanisms of crime control when imposed on a convicted offender act principally on that offender, a target the community tends to regard as the appropriate subject of interference with liberty. By contrast, the cashless bus discourages robbers by changing the behavior of bus companies and bus drivers, and inconveniencing all bus passengers. A frequent reaction to the trouble imposed on citizens by cashless buses and early store closing hours is that the wrong people are being inconvenienced. These methods generate dissatisfaction and discourage any means of preventing or diminishing crime by altering the behavior of potential victims and citizens in general. In this respect mechanisms of restraint benefit because they target those who are seen as morally appropriate subjects. Restraint shares this advantage over environmental redesign with other offender-centered techniques such as alternative punishments, treatment, supervision, and monitoring.

Moreover, restraint *controls* rather than merely *influences* the behavior of potential offenders. The risk of the drug test or a large monetary fine may persuade some potential offenders to forgo criminal opportunities, but a functioning system of restraint provides assurance that potential offenders cannot commit crimes even if they want to. The capacity to control rather than influence is the most important reason for the great and persistent popularity of incapacitation as a penal method.

The distinction between influence and control is somewhat moderated when temporal limits are imposed on incapacitation and when a quick response potential is built into monitoring systems. Crime reduction achieved by means of control lasts no longer than the last day of incarceration. Moreover, well-designed monitoring systems that can respond rapidly to initial deviance may exercise more effective control over a population in the middle term than intermittent periods of secure confinement. Nevertheless that palpable something in the capacity of a cell to hold men and women against their will both explains and to some extent justifies a public preference for restraining punishments.

Incapacitation and the Purposes of Punishment

If restraining sanctions can be distinguished from environmental defenses against crime on the basis of the element of control and the targeting of the offender for behavioral change, it is control alone that distinguishes incapacitation from the standard short list of the other purposes of criminal punishment. Deterrence and rehabilitation both include alteration of the behavior of the offender as an objective of the criminal sanction. Even the most draconian deterrent threats can achieve their purpose only when they influence the subject of the threat to choose voluntarily noncriminal conduct. To the not inconsiderable extent that effective retribution requires the subject of the sanction to acknowledge the moral significance and gravity of the offense, the achievement of the retributive purpose of criminal sanctions depends on the offender's recognition and choice. Rehabilitation seeks to alter the way in which a subject regards criminal opportunity. The rehabilitative process can be administered coercively in a prison environment, but it cannot be tested until the individual subjected to it is able to exercise free choice between crime and conformity with the law.

The reliance of incapacitation on the physical control of offenders renders it a low-technology criminal sanction and provides an avenue of assured crime reduction for even the most cynical and disenchanted observer of the penal system in the United States. Thus there exists a conveniently close fit between an emphasis on incapacitation and lack of faith in both the efficiency of criminal justice agencies and the responsiveness to the treatment of offenders. In this sense a system that places heavy emphasis on incapacitation stands in polar opposition to the sanguine view of both government and human nature that underlies the rehabilitative ideal (Allen, 1981). Incapacitation can thus be regarded as the punishment objective of last resort.

Restraint through incapacitation is also more closely tied to incarceration than any of the other recognized purposes of punishment in two respects. First, secure confinement is almost the only method of achieving total restriction of an individual's freedom of movement. There is a wide range of mechanisms available to convey social disapproval, to deter and to reeducate. However, when incapacitation is seen as the most important purpose of criminal punishment, commitment to prisons and jails must necessarily become the central technique of punishment.

Second, restraint involves placing emphasis on increasing the scale of imprisonment because incapacitation is the most transitory of punishment objectives. Threats last as long as they are credible; punishments designed to convey stigma or alter attitudes may continue to influence behavior long after the physical duration of the punishment, but crime prevention that requires physical control of its subjects lasts no longer than the doors and gates remain locked. Thus reliance on incapacitation carries the tendency not only to expand the proportion of cases punished with imprisonment but to extend the duration of prison terms.

Incapacitation and Imprisonment

There is a substantial overlap between incapacitation as an objective and imprisonment as a practice, but that overlap is by no means complete. Offenders are committed to prison and detained there for reasons other than incapacitation. Also, methods of incapacitating offenders other than imprisonment exist. This section first examines nonimprisonment strategies of incapacitation and then discusses methods of monitoring and moderating behavior—alternatives to incapacitation that lack the dimensions of complete control but may still prevent crime.

The prison and jail dominate the field of incapacitation intervention in the Western world, but they do not monopolize it. A number of methods other than incarceration serve to restrain subjects from the danger to others their free movement is believed to represent. A wider variety of stratagems of specific incapacitation are adopted when it is felt that an offender need only be restrained from one area of behavior or one instrumentality for the community to be secure. The two most frequently mentioned methods of exercising general control over an individual outside of prison and jail are house arrest and electronic surveillance.

Regimes of house arrest physically restrain subjects in their own home surroundings rather than in cells within public facilities. This strategy is suitable when an offender is not regarded as a danger at home and when relatively inexpensive methods to ensure compliance with the conditions of home confinement are available. The two principal disadvantages of home confinement as a frequent recourse for incapacitation are cost and security. If personnel need to be employed to maintain security conditions in what are, in effect, one-prisoner prisons, the costs will be large, whether borne by the government or the offender, who will thus avoid the public prison. If surveillance is relaxed or a family member becomes the custodian or agent of restraint, the system becomes much more vulnerable to the possibility of uncontrolled movement by the subject.

So-called electronic monitoring systems are recent attempts to avoid the cost-security dilemma of house arrest and to provide effective restraint by publicly employed security agents without the high cost of providing at least one security monitor per subject. Electronic monitoring systems allow a public employee at a remote location to maintain knowledge of the whereabouts of each of a large number of individuals wearing signal-sending devices attached to their bodies. Whether the system is regarded as a modification of house arrest or as a wholly new strategy, it is a widely heralded new approach in many American criminal justice systems. The appeal is that it approximates control without the consequences of full imprisonment. Electronic monitoring may be less expensive than imprisonment because the security costs of the monitoring systems do not equal the aggregate of the security and maintenance costs entailed by imprisonment.

From Control to Monitoring

Programs of intensive supervision that substitute monitoring for incapacitation are more frequently used as an alternative to imprisonment than house arrest and electronic monitoring. These are schemes that function by requiring frequent contact between the subject and a supervisor—in some cases two or three times a day, both in public offices and in the subject's home surroundings. Though such a system may provide ample early warning when a subject of supervision starts misbehaving, it does not provide the kind of physical restraint that disables subjects from offending in the community, whatever their wishes or willingness to risk the consequences.

Thus intensive supervision is one of a variety of monitoring devices that are the principal functional alternatives to incapacitation. The widely felt and sometimes expressed objection to monitoring rather than incapacitation is the lesser security that is involved when the system first has to find the subject to determine that he or she is not in compliance before any measures of restraint can be imposed.

In practice the distinction between monitoring systems and explicit physical restraint is more a matter of degree than of kind. There exist very few, if any, escape-proof prisons in the world and there are many monitoring systems with a quick response capacity. The difference between control on the one hand and monitoring on the other is symbolically important. For this reason systems of surveillance such as intensive supervision are frequently described in rhetorical terms that emphasize the capacity to control. A system that does not depend on the cooperation of offenders is one that "feels" safer to all of us.

The same choice between monitoring and controlling those who pose a threat of future criminality can be found in policies dealing with subjects who pose a threat through one particular form of behavior that may be subject to specific preventive measures. Stock and commodity brokers convicted of fraud are subjected to a measure of special incapacitation whereby they are forbidden to participate in the regulated markets that are the only places where their criminal specialties can occur. Narcotics addicts and alcohol abusers can have chemical antagonists administered to them either to destroy the effect of a desired drug or make its ingestion unbearably unpleasant.

Incapacitative loss of privileges may be imposed on physicians, lawyers, and others who have been apprehended in criminal abuse of professional privilege. Other measures of specific incapacitation may include denial of access to special security areas or the revocation of security clearances. The reason for classifying loss of access as a measure of incapacitation is the capacity of the system to deny potential offenders the opportunity to reoffend whether they want to or not. Taking away the driver's license of a drunken driver is not a measure of incapacitation because a heedless subject can drive again if he wants to do so, but taking away the same subject's automobile does involve a degree of incapacitative control.

For specific prevention the contrast between different kinds of monitoring systems may be as important as the distinction between monitoring and control. Suspending the licenses of convicted traffic offenders is a notoriously unreliable road safety measure because the defrocked driver may continue to drive for long periods of time without being detected, but designing more effective monitoring checks for specific high-risk drivers can improve the detection capacity of the system and make it competitive with incapacitative controls for all but the most overwhelmingly dangerous individuals.

Monitoring versus Control

The principal alternative to incapacitative controls, at almost every juncture in the criminal justice system, is monitoring and supervision. After arrest, supervised release is the monitoring alternative to incapacitation through preventive detention. After conviction, supervised or intensive probation is the most attractive alternative to incapacitative imprisonment. For those serving prison sentences parole authorities choose between the risk-monitoring approach of release with parole supervision and the physical control maintained by further incarceration. For drug offenders, periodic drug testing is a monitoring alternative to control by means of incarceration. At every turn, therefore, a sanctioning system chooses between monitoring and control for large numbers of offenders.

Three general points can be made about the choice between systems that monitor and those that control. First, a definite preference expressed throughout the criminal justice system will have enormous consequences for the shape and character of the system. A bias in favor of control rather than the monitoring of risk will produce a much larger prison and jail population than in a system more favorably disposed toward risk monitoring.

Second, an important consideration in the choice between monitoring and control is the perceived cost of individual cases of failure. Since risk-monitoring systems discover noncompliance *after* the fact, a critical advantage of the control alternative is the capacity to prevent the first failure. Where the harm risked through a first failure is substantial, the bias toward control rather than monitoring will be significant. When occasional failures are to be expected, and are also relatively inconsequential, a bias toward monitoring strategies makes sense. Property crimes for which the social cost of isolated instances is low, and drug addiction for which some degree of recidivism is regarded as not totally destructive of program effect, are two examples of situations in which risk-monitoring approaches would seem attractive. The possibility of life-threatening violence, on the other hand, exemplifies the sort of situation in which the chance of a first failure makes actors feel uncomfortable about risk monitoring.

Third, the quality of the risk-monitoring systems that are available as alternatives to incapacitative control should have a direct bearing on the choice between incapacitation and risk monitoring. Creating monitoring systems that detect noncompliance more quickly and give earlier warning of antisocial conduct will make the risk-monitoring approach more attractive in relation to a broad range of behaviors for which incapacitative controls are usually preferred. For this reason a special focus on improving the character and efficiency of risk-monitoring systems should be regarded as a crucially important task by those concerned to create alternatives to imprisonment. The more confidence monitoring systems can inspire, the less necessary will the incapacitative function of imprisonment seem to be.

II. Two Case Studies in Policy Choice

Concrete examples of the conflict between risk-monitoring and incapacitative responses to offenders are not hard to find. This section will address two offense categories which entail such conflict and for which the choice of alternatives involves the disposition of large numbers of offenders and has a substantial impact on total commitments to prison. The two categories, overlapping to some extent, are drug users convicted of drug or property offenses and recidivist property offenders. Together, these two categories currently account for the majority of all prison commitments in the United States and a somewhat smaller share of the U.S. prison population on any given day.

Drug Offenders

The drug-abusing criminal offender is of special significance to the student of criminal policy for multiple reasons. First, the number of persons arrested and convicted of drug offenses is much more variable over time, as policies change, than is the population arrested for other serious crimes. The number of drug arrests fluctuate up and down far more dramatically than any other offense category (Zimring and Hawkins, 1991:134–36). Lately, the level of drug arrests has been ascendant and the proportion of prisoners being punished for drug crimes has increased more rapidly since the mid-1980s than ever before in United States history. In California, as the total number of prisoners increased *fourfold,* the number serving sentences for drug offenses increased *fifteenfold* in twelve years. By 1991 commitment for drug offenses accounted for one-quarter of all California prisoners. There were more persons in California prisons as a result of drug convictions alone in 1991 than there were total prisoners at the beginning of 1980 (Zimring and Hawkins, 1992:32).

When the total number of persons convicted of drug crimes in California is augmented by the many thousands for whom drug dependency was the reason they committed other offenses, close to one-third of the prison population has a significant relation to illicit drugs. This is not atypical of other states in the United States with major metropolitan areas (U.S. Department of Justice, Bureau of Justice Statistics, 1993:1).

A further distinguishing characteristic of the drug offender as a policy problem is that therapeutic intervention for dependent offenders is a policy option that has both political and clinical support. While the imprisonment of drug offenders received substantial political and judicial support in the late 1980s, so too did a combination of community-based drug treatment and risk monitoring through the use of chemical tests designed to detect recent drug usage. The popularity of what has come to be called Treatment Alternatives for Street Crime (TASC) has been an important counterweight to widely heralded declarations of loss of faith in rehabilitation for criminal offenders. Drug treatment has a better reputation in criminal justice courts than any other therapeutic intervention for adults.

The two policy packages in competition for the middle-level drug or drug-related offenders are: relatively short imprisonment terms, perhaps coupled with after-care programs that stress drug monitoring, versus a package of community-based drug treatment with a long period of drug monitoring and with treatment repetition on detection of drug use during the monitoring program. Each alternative carries a threat of commitment to prison in the case of repetitive failure in drug monitoring. The legal justification for reimprisonment in the case of parole would be violation of parole conditions, and in the case of community-based treatment it would be violation of the probation status that universally accompanies court commitment to treatment after criminal conviction. When the treatment is a pretrial diversion from the criminal justice system, the threat of further sanction on failure comes from reviving the criminal prosecution.

One distinctive feature of the competition between community-based treatment and imprisonment for drug-abusing offenders is that each alternative promises a type of crime prevention that the other cannot deliver. Imprisonment promises the physical control that is its universal comparative advantage. Drug treatment hopes to reduce the criminal propensity of those of its subjects that it renders non-drug-dependent. Though such treatment successes will constitute only a fraction of the total treated population, the duration of treatment effects can be greater than the period of control provided by imprisonment.

There is reason to believe that combining imprisonment with drug treatment will not produce a best-of-both-worlds combination of short-term control and long-term treatment effects. As Morris and Tonry put it:

> The prison environment is far from a suitable setting for drug treatment programs or for psychological treatments . . . Intermediate punishments with conditions of treatment would therefore seem particularly appropriate for addicted criminals, provided sufficient control of their behavior can be built into those programs to satisfy legitimate community anxieties. (1990:188)

It appears that community-based treatment programs carry a larger promise of success than drug therapy inside prison.

If community-based treatment has a higher success rate than prison-based treatment, putting the offender into prison increases the amount of crime prevented by control but reduces the degree of crime prevention to be expected from successful drug treatment. Avoiding prison in favor of treatment maximizes the prevention associated with treatment but sacrifices the preventive benefits of control. With crime prevention on both sides of the equation this is a very different competition from the usual imprisonment versus alternatives-to-imprisonment comparison, where the control benefits of imprisonment are contrasted with the lower costs, lesser intrusiveness and other values of alternatives to imprisonment but where the alternatives carry no promise of any unique crime preventive benefit.

There is one other sense in which the debate about drug dependent offenders during the 1980s and early 1990s differed from the discussion of criminal justice policy toward other felons. Incapacitation through restraint seemed to be less important as a reason for imprisonment for drug offenders than for most other offenders. No claims were made that locking up drug offenders reduced the amount of drug crime occurring in the community as a result of restraint. Rhetorical appeals favoring the imprisonment of drug criminals tended to emphasize deterrent and retributive considerations.

This might mean that the imprisonment of drug offenders would be easier to displace in favor of community treatment. However, the lack of emphasis on incapacitation in drug crimes also seems to refute the hypothesis that a demand for incapacitation was necessary before the use of imprisonment would expand significantly for an offender group. The sanctions for drug offenders expanded more rapidly than for any other major offender group between 1985 and 1990 (U.S. Department of Justice, Bureau of Justice Statistics, 1993:8). That such expansion could take place without a major emphasis on incapacitation should lead to some restraint in the imputation of a causal role for incapacitative motives in the expansion of the prison system.

As the proportion of all prisoners sentenced for drug crimes has expanded, the debate about the appropriate punishment for such offenders has heated up. Pressure to divert drug interdiction and law enforcement resources to drug treatment programs is a consistent feature of the drug policy debate of the early 1990s. Rhetorically the existence of a rehabilitation regime in which the public expresses some confidence is a substantial advantage that advocates of alternatives to incarceration rarely have in the debate. If public anxiety about drugs moderates in the mid-1990s, the prospect for nonincarcerative drug treatment as a program emphasis seems quite good.

Property Offenders

A large concentration of offenders in American prisons were convicted of property offenses that do not carry any direct threat of personal violence: crimes such as burglary, larceny, auto theft, and forgery. Table 8.1 shows the distribution of state prisoners by offense of commitment for 1986 and persons admitted to state prisons in 1990.

Just under one-third of the 1986 prison population was serving a sentence for a property offense not aggravated by harm to a person. Because property offense sentences are shorter than violence sentences, the property group accounts for an even higher 37 percent of all prison admissions during 1990.

Nonviolent property offenders constitute a major share of the prison population in any Western nation. Such offenders also constitute the mass of felons who are *not* sent to prison in the United States and across Europe. For any modern criminal justice system the answer to the question of where in the distribution of property offenders the line is to be drawn between imprisonment and lesser punishment will have a

Table 8.1. Percentage of Distributions of State Prison Population and Admissions by Offense Type

Offense Type	Prison Population (1986)	Prison Admissions (1990)
Property crime (including auto theft, theft, burglary, etc.)	31	25
Robbery	21	15
Other violent crime	34	32
Drugs and other	14	28
Total	100%	100%

Source: U.S. Department of Justice, Bureau of Justice Statistics.

significant impact on the scale of the prison enterprise. A modest increase or decrease in the proportion of thieves committed to prison can have a substantial impact on the size of the prison population. Table 8.2, taken from our recent study of the growth of imprisonment in California, illustrates the trend in male imprisonment by offense category for five property offenses.

As table 8.2 illustrates, the largest percentage changes in the prison population are associated with the less aggravated forms of property offense. The number of inmates committed for robbery only doubles over a decade in which the overall prison population quadruples. The number of inmates committed for burglary increased by 335 percent during the same period, keeping pace with the overall general trend. Prisoners committed for motor vehicle theft increased 500 percent over the same decade and those committed for theft by 565 percent.

Incapacitation is a central concern in decisions about policy toward property offenders, and enthusiasm for the incapacitative benefits of imprisoning property offenders seem to have played an important role in the prison expansion of the 1980s. While much of the political rhetoric of incapacitation uses the imagery of violent crime, most sophisticated observers of the criminal justice system would acknowledge that the debate about prison expansion in the 1980s and the 1990s is about burglars, automobile thieves, and minor property offenders, as well as the drug offenders previously discussed.

The debate about policy for property offenders is nowhere near as two-sided as the debate about responses to drug offenders. The imprisonment versus treatment dialogue regarding drug offenders concerns two high-profile social institutions. Each alternative has a constituency in the community. Dominated by the prospect of imprisonment, the debate about policy toward property offenders is in most significant respects a contest between those who support and those who oppose imprisonment rather than a choice between two clearly defined institutional alternatives.

The contending policy packages for repeat property offenders reflect this lack of cleavage. Both hard-line and soft-line approaches to repetitive property felons rely on prison as a sanction of ultimate resort, but the soft-line approach temporizes more extensively, frequently involving probation together with whatever intermediate sanctions are available in particular jurisdictions.

A typical hard-line approach in the United States uses imprisonment earlier in a property offender's career but usually not as a first resort. Only residential burglary puts a first-conviction property felon at risk of imprisonment in a typical hard-line regime. For nonaggravated theft the distinction between the hard-line and soft-line policies may be the difference between short-term imprisonment for a second offense, as opposed to a fourth.

Both hard-line and soft-line rhetoric about policy toward property offenders are dominated by propositions about imprisonment. Desert and retribution are the themes emphasized by the hard-line proponents, while the characterization of imprisonment as an unnecessary and expensive overreaction is a recurrent feature in soft-line argument. To date no single intermediate sanction such as the fine or community service has attracted a consistently enthusiastic constituency among soft-line advocates in the United States. Instead, recent critiques of hard-line policies have tended to place heavy emphasis on the financial costs of prison construction and imprisonment. Indeed, some of the rhetoric of cost and benefit discussed in chapter 7 seems to have been introduced into the debate on criminal justice and penal policy as a counter to the emphasis placed on costs by critics of imprisonment.

The extent to which current policies are successful in crime prevention by means of incapacitation is discussed in contemporary policy debates only in the most general terms. Soft-line partisans point to relatively stable crime rates as evidence that additional crime prevention from increased use of prisons and jails is minimal. The hard-line advocates

Table 8.2. Male Imprisonment by Offense Category, Property Offenses, California 1980–90

Offense	1980	1990	Numerical Change	Percentage Change*
Robbery	5,791	11,828	6,037	104
Homicide	3,879	11,569	7,690	198
Burglary	3,153	13,702	10,549	335
Motor vehicle theft	392	2,346	1,954	499
Theft	1,051	6,992	5,941	565
Forgery	392	791	399	102
All prisoners	20,608	90,327	69,719	338

*Rounded to nearest percent.

Sources: California Department of Corrections and California Bureau of Criminal Justice Statistics and Special Services; Zimring and Hawkins, 1992, p. 39.

claim that the expansion of prison populations involved large numbers of high-rate offenders and that substantial aggregate crime prevention must therefore have resulted from that expansion (Barr, 1992). The two competing rhetorical perspectives rarely engage each other. It is even rarer for specific research findings regarding the effect of the implementation of penal policies to come under sustained critical scrutiny in the public debate about the incarceration of property offenders.

Because the soft-line critique of prison population expansion does not enthusiastically support any specific program of alternatives to prison, those skeptical about that expansion lack a consensus crime prevention program that they can recommend. A typical soft-line response to an argument about the need for additional imprisonment will   stress the crime prevention potential of investing large amounts of money in alternative programs, but there is no agreement whether those programs should be criminal justice programs or some kind of social programs.

Notable for its absence in the debate about incapacitation and property crime is any reference to criminal justice and penal policies in countries other than the United States, and to the impact of those policies on crime rates. While American world leadership in crimes of violence is nowhere doubted and much of this country's crime problem is exemplary of American particularity, our house breakers, automobile thieves and shoplifters are more likely to be representative of a species to be found throughout the Western world. Indeed the chronic property offender presents a major problem in criminal sentencing almost everywhere. Comparative cross-national study is overdue.

Since the topic of incapacitation is so substantial a feature of the debate about prison expansion in the United States, observers might be tempted to conclude that whoever wins the argument about the incapacitation of property offenders will exert a major influence on whether the scale of imprisonment continues to expand, stabilizes, or perhaps contracts. Any such conclusion would be premature on present evidence. Those who hope for rationality in government policy tend to believe that there is a close linkage between debates about the purpose of imprisonment in a society and decisions about the size and nature of its prison system (see Sherman and Hawkins, 1981:48ff).

However, so little is known about how decisions are made about the appropriate size of a prison system that an objective approach to the question of the influence that debates about purposes or prison size actually have on practice is warranted. Prison systems that share the same formal jurisprudence of prison sentencing vary widely in size and degree of punitiveness. Two systems may have an identical commitment to both desert and incapacitation, but one can send its burglars to prison on a second conviction while the other waits for a fourth offense. Moreover, declarations about the purposes of imprisonment are commonly ex post facto justifications for prison policies that have already

been decided and implemented rather than being the approved prior
motives for opening or closing prisons.

If the influence of public attitudes on the officially accepted purposes
of imprisonment and on the size of the prison system is an open question, it is also an important question for those who are interested in the
role of incapacitation in penal policy. If the size of a prison system is
determined quite independently of its dominant purpose or purposes,
the influence of the idea of incapacitation on the implementation of the
criminal law may be decisive in regard to *which* offenders are sent to
prison but not in relation to *how many*. In those circumstances an emphasis on incapacitation might result in leniency for some classes of offenders, as more attractive candidates for incapacitation crowd out those who
might be imprisoned for other reasons. If on the other hand a heavy
emphasis on incapacitation leads to a substantial expansion in prison
population, the need to incapacitate may function as a supplementary
justification for imprisonment, making it no less likely that offenders will
be incarcerated for other purposes. Comprehending the probable effects
of incapacitation on penal policy thus requires attention to two competing models of how decisions are taken about the scale of imprisonment
in particular political units.

III.   Incapacitation and the Limits of Imprisonment

Surprisingly little is known about the factors that influence the size of a
prison and jail system, or the degree to which the scale of imprisonment
may be expected to change over time. What is not known about this
subject could fill a book and indeed quite recently did (Zimring and
Hawkins, 1991). For present purposes we will attempt to compensate
for the lack of available information by positing two competing models
of the social determinants of imprisonment; based on theories of fixed
and of variable imprisonment scale. We will then examine some of the
implications of each model for the likely influence of notions about
incapacitation on penal policy.

Fixed versus Variable Models

One need not subscribe to a wholly arbitrary explanation of how prison
systems come to be built to believe that the existing capacity to imprison
in a political unit is unlikely to vary substantially over time in all but
unusual circumstances. Any number of theories of original causation can
be consistent with the hypothesis that the holding capacity of a prison
and jail system will vary only modestly over long periods of time. Analysis
of historical data from the United States and some other countries led
Alfred Blumstein and his associates to propose a theory of "the stability

of punishment" during the 1970s (Blumstein and Cohen, 1973; Blumstein, Cohen, and Nagin, 1976; Blumstein and Moitra, 1979). Whereas the explosive growth in prison population in most U.S. states in recent years has conclusively disproved "the stability of punishment" as a universal rule, the general pattern found among Western nations is for countries to vary widely amongst each other in prison population but for their relative ranking not to change significantly over time.

When the factors that limit the capacity of a government to imprison operate independently of changes in fashion in penal purposes and create long-term fixed limits, changing priorities within the criminal law are apt to have more influence on *which offenders* are sentenced to imprisonment and for *how long* than on the total number of those imprisoned. Under these circumstances ideas of collective incapacitation will be a relatively unimportant influence on penal policy and practice, but the notion of selective incapacitation may well assume considerable importance. A criminal justice system operating within fixed limits will search for those of its convicted offenders with the greatest propensity to reoffend and will try to avoid using scarce confinement space for lower-risk subjects. Studies of subpopulations of offenders will hold promise of policy application, whereas studies that attempt to derive average levels of crime avoidance for an entire population of imprisoned offenders will have less obvious relevance to imprisonment policy.

If the limits of imprisonment are more flexible over time, data on the possible benefits of collective incapacitation may play a more direct role in the formation of penal policy. Also, if prison population can expand or contract significantly as a result of changes in attitudes toward the purposes of imprisonment, shifts in the value placed on collective incapacitation may have substantial effect on aggregate levels of imprisonment.

As we have already noted, it is by no means clear that changes in public attitudes toward incapacitation played a major role in the expansion of imprisonment in the United States since the late 1970s. Yet it is worthwhile to consider the appropriate influence that particular findings about incapacitation might have on the desired size of a prison system. We have argued that when prison populations are relatively fixed, data about collective incapacitation results are of limited value in determining prison policy, whereas identifying subgroups of high-risk offenders is a research task with much more direct policy application. Are findings about collective incapacitation more directly relevant to policy when levels of imprisonment can vary by a wide margin?

The expected crime prevention benefits that can come from either imprisoning or releasing large groups of criminal offenders are directly relevant to policy decisions in such circumstances. However, the important group for policy analysis here should not be the average offender in the prison system but the offender on the margin between imprisonment and nonimprisonment sanctions. If the processes that produce decisions about imprisonment are selective, as that term was defined in chapter 3, the expected crime savings from incarcerating offenders at the margin of imprisonment policy will be quite different from those one would expect from the incarceration of the average imprisoned offender.

Because that is the case, the process of identifying likely subjects of a policy change and attempting to measure responses to such a change will more closely resemble the analysis of selective incapacitation initiatives than the analysis of systemwide averages. The benefits derived from imprisoning the average incarcerated offender would be of policy significance only in jurisdictions that are nonselective in the use of imprisonment or that might be considering the abolition of their prison systems.

The Myth of the Optimum

Can the careful analysis of data on marginal offenders tell us what level of prison population would be optimum from the standpoint of incapacitation policy? If the reader agrees with our analysis of monetized measures of the cost of crime and law enforcement in chapter 7, there is no economic point of optimal incarceration to be identified. The problem with the notion of an optimal level of incarceration is not merely an absence of data, it is an absence of meaning.

At most, research can provide us with estimates of the number and type of offenses restrained by particular levels of imprisonment. These data can then be compared with information on the public and private costs of other crime prevention programs, and some rough conception of appropriate prison size may emerge from the political process. However, as long as incapacitation benefits do not fall to zero, incapacitation will function more as an open-ended justification for imprisonment that will be part of the political dialectic that determines the size of a prison system.

Thus, even if the scale of imprisonment is a variable value and even if it varies in relation to public sentiment about the various purposes of imprisonment, these processes cannot be expected to produce any optimization in the determination of levels of imprisonment. This is important because it suggests that fewer opportunity costs occur when governments operate with prison systems that have fixed and essentially arbitrary population limits. If the processes that determine prison scale under conditions of variable capacity are arbitrary and politically actuated, there is no a priori reason to suppose that prison capacity levels in such systems operate any better to serve the public interest than in fixed-capacity systems.

This indeterminacy is particularly comforting in view of the substantial indications we have seen that prison capacities are relatively fixed in most political systems most of the time. If a criminal justice system responding to relatively fixed levels of imprisonment can protect its population from criminal harm as effectively as a system subject to greater degrees of variation, the lack of variation in prison scale that is characteristic of much of modern government should not be regarded as a cause for regret.

The Future of an Idea

What can be said about the future of incapacitation as a penal fashion? Incapacitation has always been an important function of incarceration even when the concept was ignored in the criminological literature. We think that the current heightened interest in the topic is unlikely to abate. Instead restraint is likely to remain a more visibly important function of prisons and jails than has been the case during most of the twentieth century. Though the subject of incapacitation is unlikely to stimulate an outpouring of scholarly concern, we would also anticipate a steady stream of empirical and theoretical work to be a consistent feature of scholarship in criminal justice.

Undoubtedly, the kind of scholarship produced on the topic will be influenced by fluctuating levels of enthusiasm in those political and academic circles so that trends in the ways of thinking about incapacitation are not only interesting in their own right but can also tell us a good deal about future scholarly involvement with the topic. Two changes in emphasis are likely during the next decade: a move from heavy emphasis on collective incapacitation to an emphasis on selective incapacitation possibilities; and a less dominant position for incapacitation vis-à-vis the other objectives of the criminal justice system.

Two reasons lead us to expect there will be a sharp increase in interest about selective incapacitation policies. First, the recent history of incapacitation literature has been characterized by an alternating focus on collective and selective approaches. Collective incapacitation has been at center stage in the United States for almost a decade and so the pendulum should soon swing back.

The second reason to expect a shift in emphasis from collective to selective incapacitation policies is the limited prospect for further expansion of the prison system in the United States. The period since the mid-1970s has seen a large and uninterrupted increase in prison population under various forms of incarceration. Collective incapacitation as a policy emphasis best functions in an environment where large variations in the numbers confined can occur. If the prospect of further expansion of prisons and jails meets increasing resistance, those wishing to increase crime control through mechanisms of restraint will increasingly be pushed toward a more efficient use of existing incarceration resources rather than a continued increase in the scale of imprisonment. If correctional policy is, as Bismarck said of all politics, "the art of the possible," the incentive to adopt selective strategies will be substantial.

The problems associated with successfully identifying high-rate offenders have not significantly diminished over the years since selective incapacitation was a major policy focus, but the opportunity to avoid those problems and still make major claims for the future crime prevention potential of prisons and jails has diminished.

We also expect the dominance of incapacitation in the dialogue about penal purposes to abate somewhat in the near future. An emphasis on incapacitation is particularly suited to periods of pessimism about the efficacy of government programs and the capacity of individuals to undergo positive change. It is in this context that we have referred to incapacitation as a penal purpose of last resort. Historical evidence demonstrates that attitudes regarding both the malleability of offenders and the capacity of government programs to achieve constructive results are cyclical (Allen, 1981). Any increase in public confidence about the prospect of using a wider variety of crime control strategies is likely to reduce the relative emphasis on incapacitation from its recent levels in criminal justice administration. Some diminution in the dominance of incapacitation in popular thought would represent an elevation of public mood about crime control.

We expect public attachment to the idea of incapacitation to fall to a lower level than has been evident in recent years but remain at one considerably higher than most of modern history. A moderation in the recent emphasis on restraint as a penal purpose can benefit both policy and research. Worthwhile scholarship and intelligent policymaking are most likely to flourish in a social atmosphere in which incapacitation is recognized as an important, but by no means exclusive, means of social defense against serious crime.

Appendix

Estimated Offenses Avoided per Additional Year of Confinement,
California, 1981–90

Year	Method 1	Method 2	Method 3	Method 4
Larceny				
1981	0.99	−1.86	0.99	−1.86
1982	0.68	−3.66	0.45	−4.92
1983	3.98	2.19	10.33	13.84
1984	3.75	1.60	2.90	−0.02
1985	2.65	0.79	−0.90	−1.68
1986	2.08	0.68	0.03	0.26
1987	2.27	1.87	2.89	6.68
1988	1.82	0.69	−1.25	−5.99
1989	1.44	0.83	−1.03	1.43
1990	1.74	1.48	3.66	5.94
Assault				
1981	1.01	−1.65	1.01	−1.65
1982	0.78	0.20	0.61	1.59
1983	0.60	0.17	0.23	0.11
1984	0.41	0.28	−0.19	0.56
1985	0.28	0.21	−0.12	−0.04
1986	−0.42	−0.39	−2.68	−2.35
1987	−0.51	−0.51	−0.84	−0.98
1988	−0.47	−0.35	−0.14	0.87
1989	−0.48	−0.34	−0.50	−0.18
1990	−0.49	−0.18	−0.41	1.51
Burglary				
1981	2.13	4.30	2.13	4.30
1982	4.12	2.32	5.44	1.14
1983	4.74	4.13	5.79	7.44
1984	4.33	2.24	2.93	−2.25
1985	3.50	2.12	0.67	1.44
1986	2.76	1.26	0.13	−1.34
1987	2.98	1.72	3.60	3.54
1988	2.88	1.73	1.90	1.64
1989	2.61	1.62	0.57	0.69
1990	2.56	1.65	1.58	1.50

Year	Method 1	Method 2	Method 3	Method 4
Auto Theft				
1981	2.38	−2.92	2.38	−2.92
1982	1.06	−2.22	0.13	−1.95
1983	1.07	−1.67	1.04	−0.96
1984	0.81	−1.33	0.05	−0.37
1985	0.37	−0.82	−1.01	1.48
1986	−0.13	−0.83	−1.75	−0.58
1987	−0.37	−0.37	−1.39	2.58
1988	−0.70	−0.46	−2.60	−0.71
1989	−0.87	−0.53	−1.78	−0.69
1990	−0.75	−0.49	0.35	−0.19
Robbery				
1981	−0.27	−0.80	−0.27	−0.80
1982	0.12	−0.62	0.39	−0.54
1983	0.40	−0.46	0.94	−0.25
1984	0.40	0.09	0.39	1.87
1985	0.31	0.26	−0.001	0.73
1986	0.17	0.17	−0.29	−0.14
1987	0.30	0.37	0.84	1.17
1988	0.26	0.36	−0.07	0.16
1989	0.14	0.18	−0.56	−0.81
1990	0.01	−0.03	−1.00	−1.55
Rape				
1981	0.06	−0.03	0.06	−0.03
1982	0.10	−0.03	0.14	−0.04
1983	0.10	0.08	0.08	0.30
1984	0.09	0.11	0.07	0.19
1985	0.08	0.12	0.05	0.11
1986	0.06	0.11	−0.03	0.03
1987	0.05	0.11	0.02	0.08
1988	0.05	0.09	0.05	−0.01
1989	0.05	0.08	0.01	0.004
1990	0.04	0.08	−0.03	0.04
Murder				
1981	0.050	−0.015	0.050	−0.015
1982	0.048	0.031	0.045	0.064
1983	0.040	0.018	0.023	−0.004
1984	0.029	0.015	−0.004	0.005
1985	0.023	0.010	0.002	−0.005
1986	0.014	0.006	−0.013	−0.007
1987	0.014	0.006	0.013	0.007
1988	0.013	0.006	0.004	0.004
1989	0.010	0.012	−0.009	0.039
1990	0.007	0.011	−0.023	−0.006

References

Abell, Richard B. (1989). "Beyond Willie Horton: The Battle of the Prison Bulge." *Policy Review* (winter):32–35.

Allen, Francis A. (1981). *The Decline of the Rehabilitative Ideal: Penal Policy and Social Purpose.* New Haven, CT: Yale University Press.

American Criminal Law Review (1979). "A Closer Look at Habitual Criminal Statutes: *Brown v. Parratt* and *Martin v. Parratt,* A Case Study of the Nebraska Law." *American Criminal Law Review* 16:275–316.

Austin, James. (1986). "Using Early Release to Relieve Prison Crowding." *Crime and Delinquency* 32:404–502.

Barr, William P. (1992). Speech to California District Attorney's Association. *Federal Sentencing Reporter* 4(6):345–46.

Bassett, Margery (1943). "Newgate Prison in the Middle Ages." *Speculum* 18:233.

Becker, Gary S. (1968). "Crime and Punishment: An Economic Approach." *Journal of Political Economy* 76:169–217.

Bentham, Jeremy ([1789] 1841). "Introduction to the Principles of Morals and Legislation." *The Works of Jeremy Bentham,* Vol. 1, pp. 1–154. London: Simpkin, Marshall.

_____ ([1802] 1843). "Panopticon versus New South Wales." In J. Bowring (ed.), *The Works of Jeremy Bentham,* Vol. 4, pp. 173–248. London: Simpkin, Marshall.

Blue Ribbon Commission on Inmate Population Management (1990). *Final Report.* Sacramento, CA: Blue Ribbon Commission.

Blumstein, Alfred (1983). "Incapacitation." In Sanford H. Kadish (ed.), *Encyclopedia of Crime and Justice,* Vol. 3, pp. 873–880. New York: The Free Press.

_____ (1984). "Planning for Future Prison Needs." *University of Illinois Law Review* 8:1–26.

_____ (ed.) (1988). *Report of the NRC Working Group on Crime and Violence.* Mimeograph. Washington, DC: National Research Council.

Blumstein, Alfred, José A. Canela-Cacho, and Jacqueline Cohen (1993). "Filtered Sampling from Populations with Heterogeneous Event Frequencies." *Management Science* 39(7):886–99.

Blumstein, Alfred and Jacqueline Cohen (1973). "A Theory of the Stability of Punishment." *Journal of Criminal Law and Criminology* 64:198–207.

_____ (1979). "Estimation of Individual Crime Rates from Arrest Records." *Journal of Criminal Law and Criminology* 70:561–85.

Blumstein, Alfred, Jacqueline Cohen, and Daniel Nagin (1976). "The Dynamics of a Homeostatic Punishment Process." *Journal of Criminal Law and Criminology* 67:317–34.

_____ (eds.) (1978). *Deterrence and Incapacitation: Estimating the Effects of Criminal Sanctions on Crime Rates.* Washington, DC: National Academy of Sciences.

Blumstein, Alfred, Jacqueline Cohen, Jeffrey A. Roth, and Christy A. Visher (eds.) (1986). *Criminal Careers and "Career Criminals,"* Vol. 1. Washington, DC: National Research Council, National Academy Press.

Blumstein, Alfred, and Soumyo Moitra (1979). "An Analysis of Time Series of the Imprisonment Rate in the States of the United States: A Further Test of the Stability of Punishment Hypothesis." *Journal of Criminal Law and Criminology* 70:376–90.

Boland, Barbara (1978). "Incapacitation of the Dangerous Offender: The Arithmetic Is Not So Simple." *Journal of Crime and Delinquency* 15:126–29.

Campbell, Donald T. (1971). "Legal Reforms as Experiments." *Journal of Legal Education* 23(1):217–39.

Chaiken, Jan M., and Marcia R. Chaiken (1982). *Varieties of Criminal Behavior.* Rand Report R-2814-NIJ. Santa Monica, CA: Rand Corporation.

Clarke, Stevens H. (1974). "Getting 'Em Out of Circulation: Does Incarceration of Juvenile Offenders Reduce Crime?" *Journal of Criminal Law and Criminology* 65(4):528–35.

Cohen, Jacqueline (1978). "The Incapacitative Effect of Imprisonment: A Critical Review of the Literature." In Alfred Blumstein, Jacqueline Cohen, and Daniel Nagin (eds.), *Deterrence and Incapacitation: Estimating the Effects of Criminal Sanctions on Crime Rates.* Washington, DC: National Academy of Sciences.

_____ (1983). "Incapacitation as a Strategy for Crime Control: Possibilities and Pitfalls." In Norval Morris and Michael Tonry (eds.), *Crime and Justice: An Annual Review of Research,* Vol. 5, pp. 1–84. Chicago: University of Chicago Press.

_____ (1984). "Selective Incapacitation: An Assessment." *University of Illinois Law Review* 1984(2):253–90.

_____ (1986). "Research on Criminal Careers: Individual Frequency Rates and Offense Seriousness." In Alfred Blumstein et al. (eds.), *Criminal Careers and "Career Criminals,"* Vol. 1. Washington, DC: National Research Council, National Academy of Sciences.

Cohen, Jacqueline, and José A. Canela-Cacho (1994). "Incapacitation and Violent Crime." In Albert J. Reiss and Jeffrey Roth (eds.), *Understanding and Preventing Violence,* Vol. 4, pp. 296–388. Washington, DC: National Academy of Sciences.

Cohen, Mark A. (1988). "Pain, Suffering, and Jury Awards: A Study of the Cost of Crime to Victims." *Law and Society Review* 22(3):537–55.

Conard, Alfred (1964). *Automobile Accident Compensation*. Ann Arbor: University of Michigan Press.

Conrad, Frederick A. (1942). "Statistics in the Analysis of Social Problems." *Sociology and Social Research* 26:538–49.

Cook, Philip J. (1983). "Costs of Crime." In Sanford Kadish (ed.), *Encyclopedia of Crime and Justice*, Vol. 1, 373–78.

Cressey, Donald (1958). "Achievement of an Unstated Organizational Goal: An Observation on Prisons." *Pacific Sociological Review* 1:43.

Cross, Rupert (1971). *Punishment, Prison and the Public*. London: Stevens and Sons.

Currie, Elliott (1982). "Crime and Ideology." *Working Papers* 9(3):26–35.

Dilulio, John J. (1989). "Punishing Smarter: Penal Reforms for the 1990s." *The Brookings Review* (summer):3–12.

———— (1990). *Crime and Punishment in Wisconsin: A Survey of Prisoners and an Analysis of the Net Benefit of Imprisonment in Wisconsin*. Milwaukee: Wisconsin Policy Research Institute.

Dilulio, John J., and Anne Morrison Piehl (1991). "Does Prison Pay? The Stormy National Debate over the Cost-Effectiveness of Imprisonment." *The Brookings Review* (Fall):28–35.

Ekland-Olson, Sheldon, and William R. Kelly (1993). *Justice Under Pressure: A Comparison of Recidivism Patterns Among Four Successive Parolee Cohorts*. New York: Springer-Verlag.

English, Kim, and Mary J. Mande (1992). *Measuring Crime Rates of Prisoners*. Denver: Colorado Department of Public Safety, Division of Criminal Justice, Office of Research and Statistics.

Feinberg, Joel (1970). *Doing and Deserving*. Princeton, NJ: Princeton University Press.

Flanagan, Timothy J. and Kathleen Maguire (eds.) (1991). *Sourcebook of Criminal Justice Statistics 1991*. U.S. Department of Justice, Bureau of Justice Statistics. Washington, DC: U.S. Government Printing Office.

Ford, Gerald (1975). "Remarks of the President at Yale University Sesquicentennial Convocation Dinner, April 25, 1975." In *U.S. President, Public Papers of the Presidents of the United States, Gerald R. Ford, 1975*, Book 1, pp. 587–94. Washington, DC: U.S. Government Printing Office.

Fox, Lionel W. (1952). *The English Prison and Borstal Systems*. London: Routledge and Kegan Paul.

Frankel, Marvin E. (1973). *Criminal Sentences: Law Without Order*. New York: Hill and Wang.

Gillespie, Robert N. (1976). "The Cost of Crime in Illinois." *Illinois Business Review* 33:6–8.

Gottfredson, Michael, and Travis Hirschi (1986). "The True Value of Lamda Would Appear to be Zero: An Essay on Career Criminals, Criminal Careers, Selective Incapacitation, Cohort Studies, and Related Topics." *Criminology* 24(2):213–34.

Gramm, Senator Phil (1993). "Drugs, Crime and Punishment: Don't Let Judges Set Crooks Free." *The New York Times*, July 8, p. 19.

Greenberg, David F. (1975). "The Incapacitative Effects of Imprisonment: Some Estimates." *Law and Society* 9(4):541–586.

―――― (1991). "Modeling Criminal Careers." *Criminology* 29(1):17–46.

Greenwood, Peter W., with Allan Abrahamse (1982). *Selective Incapacitation: Report Prepared for the National Institute of Justice.* Santa Monica, CA: Rand Corporation.

Greenwood, Peter W., and Susan Turner (1987). *Selective Incapacitation Revisited: Why the High-Rate Offenders Are Hard to Predict.* Santa Monica, CA: Rand Corporation.

Grünhut, Max (1948). *Penal Reform: A Comparative Study.* Oxford: Oxford University Press.

Hawkins, E. R., and Willard Waller (1936). "Critical Notes on the Cost of Crime." *Journal of Criminal Law and Criminology* 26:679–94.

Horney, Julie, and Ineke Haen Marshall (1991). "Measuring Lambda Through Self-Reports." *Criminology* 29(3):471–95.

Johnson, Perry M. (1978). "The Role of Penal Quarantine in Reducing Violent Crime." *Crime and Delinquency* 24:465–85.

Langan, Patrick A. (1991). "America's Soaring Prison Population." *Science,* 251:1568–73.

Leslie, Shane (1938). *Sir Evelyn Ruggles-Brise.* London: John Murray.

Lewis, Donald E. (1986). "The General Deterrent Effect of Longer Sentences." *British Journal of Criminology* 26:47–62.

Lipton, Douglas, Robert Martinson, and Judith Wilks (1975). *The Effectiveness of Correctional Treatment: A Survey of Treatment Evaluation Studies.* New York: Praeger.

Mandeville, Bernard ([1725] 1964). *An Enquiry into the Causes of the Frequent Executions at Tyburn.* Los Angeles, CA: Augustan Reprint Society.

Marquis, Kent H., and Patricia A. Ebener (1981). *Quality of Prisoner Self-Reports: Arrest and Conviction Response Errors.* Santa Monica, CA: Rand Corporation.

Marsh, Jeffrey, and Max Singer (1972). "Soft Statistics and Hard Questions." Mimeographed discussion paper HI-1712-DP. Croton-on-Hudson, NY: Hudson Institute.

Martin, J. P. (1965). "The Cost of Crime: Some Research Problems." *International Review of Criminal Policy* 23:57–63.

Martin, J. P. and J. Bradley (1964). "Design of a Study of the Cost of Crime." *British Journal of Criminology* IV(6):591–603.

Martinson, Robert (1974). "What Works? Questions and Answers about Prison Reform." *Public Interest* 35:22–54.

McConville, Sean (1981). *A History of English Prison Administration,* Vol. 1, *1750–1877.* London: Routledge and Kegan Paul.

Messinger, Sheldon L. (1982). Review of *Imprisonment in America: Choosing the Future* by Michael Sherman and Gordon Hawkins. *American Bar Foundation Research Journal* (fall):1197–1201.

Miranne, Alfred C., and Michael R. Geerken (1991). "The New Orleans Inmate Survey: A Test of Greenwood's Predictive Scale." *Criminology* 29(3): 497–518.

Moore, Mark H. (1977). *Buy and Bust.* Lexington, MA: Heath Lexington.

Moore, Mark H., Susan Estrich, and Daniel McGillis, with William Spelman (1983). *Dealing with Dangerous Offenders.* Cambridge, MA: Harvard University, Kennedy School of Government.

———— (1985). *Dangerous Offenders: The Elusive Target of Justice.* Cambridge, MA: Harvard University Press.

Morris, Norval (1951). *The Habitual Criminal.* Cambridge, MA: Harvard University Press.

———— (1974). *The Future of Imprisonment.* Chicago: University of Chicago Press.

———— (1976). "Punishment, Desert, and Rehabilitation." In *Equal Justice Under Law,* Bicentennial Lecture Series. Washington, DC: U.S. Department of Justice.

———— (1984). "On 'Dangerousness' in the Judicial Process." *Rec A.B. City N.Y.* 39:102–28.

Morris, Norval, and Marc Miller (1985). "Prediction of Dangerousness." In Norval Morris and Michael Tonry (eds.), *Crime and Justice: An Annual Review of Research,* Vol. 6, pp. 1–50. Chicago: University of Chicago Press.

Morris, Norval, and Michael Tonry (1990). *Between Prison and Probation: Intermediate Punishments in a Rational Sentencing System.* New York: Oxford University Press.

Nagin, Daniel S., and Kenneth C. Land (1993). "Age, Criminal Careers and Population Heterogeneity: Specification and Estimation of a Nonparametric Mixed Poisson Model." *Criminology* 31(3):327–62.

National Commission on Law Observance and Enforcement (Wickersham Commission) (1931). Report, Vol. 12, *The Cost of Crime.* Washington, DC: U.S. Government Printing Office.

National Council on Crime and Delinquency (1973). "The Nondangerous Offender Should Not Be Imprisoned: A Policy Statement." *Crime and Delinquency* 19(4):449–56.

Nokes, Peter L. (1967). *The Professional Task in Welfare Practice.* London: Routledge and Kegan Paul.

Palmer, Jan, and John Salimbene (1978). "The Incapacitation of the Dangerous Offender: A Second Look." *Journal of Crime and Delinquency* 15:130–34.

Petersilia, Joan (1978). "The Validity of Criminality Data Derived from Personal Interviews." In Charles F. Welford (ed.), *Quantitative Studies in Criminology.* Beverly Hills, CA: Sage.

———— (1980). "Criminal Career Research: A Review of Recent Evidence." In Norval Morris and Michael Tonry (eds.), *Crime and Justice. An Annual Review of Research,* Vol. 2. Chicago: University of Chicago Press.

Petersilia, Joan, and Peter W. Greenwood (1978). "Mandatory Prison Sentences: Their Projected Effects on Crime and Prison Populations." *Journal of Criminal Law and Criminology* 69:604–615.

Petersilia, Joan, Peter W. Greenwood, and Marvin Lavin (1977). *Criminal Careers of Habitual Felons.* Santa Monica, CA: Rand Corporation.

Peterson, Mark A., and Harriet B. Braiker (1980). *Doing Crime: A Survey of California Prison Inmates,* Report R-2200-DOJ. Santa Monica, CA: Rand Corporation.

Peterson, Mark A., and Harriet B. Braiker, with Suzanne M. Polich (1981). *Who Commits Crimes: A Survey of Prison Inmates.* Cambridge, MA: Oelgeschlager, Gunn, and Hain.

Peterson, M., J. Chaiken, P. Ebener, and P. Honig (1982). *Survey of Prison and Jail Inmates: Background and Methods.* Santa Monica, CA: Rand Corporation.

President's Commission on Law Enforcement and Administration of Justice (1967). *The Challenge of Crime in a Free Society.* Washington, DC: U.S. Government Printing Office.

Pugh, Ralph Bernard (1968). *Imprisonment in Medieval England.* Cambridge: Cambridge University Press.

Reiss, Albert J. (1986). "Co-offending Influences on Criminal Careers." In Alfred Blumstein, Jacqueline Cohen, Jeffrey A. Roth, and Christy A. Visher (eds.), *Criminal Careers and "Career Criminals,"* Vol. 2, pp. 121–60. Washington, DC: National Research Council, National Academy Press.

Rice, A. K. (1958). *Productivity and Social Organization.* London: Tavistock.

Rolph, John E., Jan M. Chaiken, and Robert L. Houchens (1981). *Methods for Estimating Crime Rates of Individuals.* Santa Monica, CA: Rand Corporation.

Ruggles-Brise, Evelyn (1924). *Prison Reform at Home and Abroad.* London: Macmillan.

Sechrest, L., S. O. White, and E. Brown (eds.) (1979). *The Rehabilitation of Criminal Offenders: Problems and Prospects.* Washington, DC: National Academy of Sciences.

Sharpe, J. A. (1990). *Judicial Punishment in England.* London and Boston: Faber and Faber.

Sherman, Michael, and Gordon Hawkins (1981). *Imprisonment in America: Choosing the Future.* Chicago: University of Chicago Press.

Shinnar, Shlomo, and Reuel Shinnar (1975). "The Effects of the Criminal Justice System on the Control of Crime: A Quantitative Approach." *Law and Society Review* 9:581–611.

Smith, Eugene (1901). *The Cost of Crime: Report Prepared for the International Prison Commission.* Washington, DC: U.S. Government Printing Office.

Spelman, William (1984). "A Sensitivity Analysis of the Rand Inmate Surveys." Paper presented at 1984 meeting of the American Society of Criminology, Cincinnati, Ohio. Washington, DC: Police Executive Research Forum.

——— (1986). *The Depth of a Dangerous Temptation: Another Look at Selective Incapacitation.* Prepared for National Institute of Justice, U.S. Department of Justice, February 1986.

——— (1994). *Criminal Incapacitation,* New York: Plenum Press.

Thomas, J. E. (1972). *The English Prison Officer Since 1850: A Study in Conflict.* London: Routledge and Kegan Paul.

Tonry, Michael and Franklin E. Zimring (eds.) (1983). *Reform and Punishment: Essays on Criminal Sentencing.* Chicago: University of Chicago Press (1983).

U.S. Department of Justice, Office of the Attorney General (1982). *Combating Violent Crime: Twenty-Four Recommendations to Strengthen Criminal Justice.* Washington, DC: U.S. Department of Justice.

U.S. Department of Justice, Bureau of Prisons (1947). *Federal Bureau of Prisons: Annual Report.* Washington, DC: U.S. Department of Justice.

U.S. Department of Justice, Bureau of Justice Statistics (1980–92). *Sourcebook of Criminal Justice Statistics.* Washington, DC: U.S. Government Printing Office.

U.S. Department of Justice, Bureau of Justice Statistics (1993). *Prisoners in 1992.* Washington, DC: U.S. Department of Justice.

U.S. Department of Justice, Federal Bureau of Investigation (1980, 1990, 1992). *FBI Uniform Crime Reports.* Washington, DC: U.S. Government Printing Office.

U.S. Sentencing Commission (1987). *Sentencing Guidelines and Policy Statements.* Washington, DC: U.S. Government Printing Office.

Van Dine, Stephan, John P. Conrad, and Simon Dinitz (1977). "The Incapacitation of the Dangerous Offender: A Statistical Experiment." *Journal of Research in Crime and Delinquency* 14:22–35.

_____ (1978). "Response to Our Critics." *Journal of Research in Crime and Delinquency* 15:135–39.

_____ (1979). *Restraining the Wicked: The Dangerous Offender Project.* Lexington, MA: Lexington Books.

von Hirsch, Andrew (1974). "Prediction of Criminal Conduct and Preventive Confinement of Convicted Persons." *Buffalo Law Review* 21(3):717–58.

_____ (1976). *Doing Justice: The Choice of Punishment.* New York: Hill and Wang.

_____ (1985). *Past or Future Crimes: Deservedness and Dangerousness in the Sentencing of Criminals.* New Brunswick, NJ: Rutgers University Press.

von Hirsch, Andrew, and Don Gottfredson (1984). "Selective Incapacitation: Some Queries on Research Design and Equity." *New York University Review of Law and Social Change* 12(1):11–51.

Wilkins, Leslie T. (1974). "Directions for Corrections." *Proceedings of the American Philosophical Society* 118:235–47.

Wilson, James Q. (1973). "If Every Criminal Knew He Would Be Punished If Caught . . . ." *New York Times Magazine,* January 28, p. 9.

_____ (1975). *Thinking about Crime.* New York: Basic Books.

Wilson, Rob (1977). "U.S. Prison Population Again Hits New High." *Corrections Magazine* 3:1.

Wolfgang, Marvin E., Robert M. Figlio, and Thorsten Sellin (1972). *Delinquency in a Birth Cohort.* Chicago: University of Chicago Press.

Zedlweski, Edwin W. (1985). "When Have We Punished Enough?" *Public Administration Review* 45:771.

_____ (1987). *Making Confinement Decisions.* Washington, DC: National Institute of Justice.

Zimring, Franklin E. (1978). "Policy Experiments in General Deterrence, 1970–1975." In Alfred Blumstein, Jacqueline Cohen, and Daniel Nagin (eds.), *Deterrence and Incapacitation: Estimating the Effects of Criminal Sanctions on Crime Rates.* Washington, DC: National Academy Press.

_____ (1981). "Kids, Groups, and Crime: Some Implications of a Well-Known Secret." *Journal of Criminal Law and Criminology* 72:867–85.

_____ (1989). "Response to Zedlewski." *Crime and Delinquency* 35:316.

_____ (1993). "Drug Treatment as a Criminal Sanction." *University of Colorado Law Review* 64:809–825.

Zimring, Franklin E., and Gordon Hawkins (1973). *Deterrence: The Legal Threat in Crime Control.* Chicago: University of Chicago Press.

Zimring, Franklin E., and Gordon Hawkins (1986). "Dangerousness and Criminal Justice." *Michigan Law Review* 85:481.

_____ (1988). "The New Mathematics of Imprisonment." *Crime and Delinquency* 34:425–36.

_____ (1991). *The Scale of Imprisonment.* Chicago: University of Chicago Press.

_____ (1992). *Prison Population and Criminal Justice Policy in California.* Berkeley, CA: Institute of Governmental Studies.

Index